# PLANNING AND CITIES

General Editor

GEORGE R. COLLINS, Columbia University

# Architecture, Cities and the Systems Approach

**Francis Ferguson**

George Braziller     New York

Published simultaneously in Canada by Doubleday Canada, Limited
All rights reserved
For information address the publisher:
*George Braziller, Inc., One Park Avenue, New York, N.Y. 10016*
Library of Congress Catalog Card Number: 74–80660
International Standard Book Number: 0–8076–0763–0, cloth
0–8076–0764–9, paper
Printed in the U.S.A.
First Printing

# Acknowledgments

Chester Rapkin's early support and encouragement of this effort is very much appreciated. It provided the initial opportunity for much of the research.

Sigurd Grava, Mohammad Qadeer and Susan Deutsch provided many interesting insights and their discussions were very useful. Since we could never agree who won the arguments they cannot be blamed for what follows. Grace Milgram and Stephen Carroll critically reviewed an early draft of some of the material and suggested useful paths for its development.

I very much appreciate the help of Betty McCloskey with the typing and Anthony Kadysewski and Elaine Parker on the drawings. Finally, the effort was supported by a research grant from the National Endowment for the Arts.

The author and publisher wish to thank the following for permission to reprint certain materials included in this book from the publications listed below:

Academic Press—for passages from *Systems and Simulation* by Dimitrius N. Chorafas (1965) p. 1.

American Society of Planning Officials—for passages from *Planning 1965* by Earl A. Levin.

Atheneum—for passages from *The Economics of Defense in the Nuclear Age* by Charles J. Hitch and Roland McKean (1967) pp. 118–119.

Atlantic-Little-Brown—for passages from *The Architecture of America* by John Burchard and Albert Bush-Brown (1961) pp. 20–21.

Basic Books—for passages from *Toward a National Urban Policy* edited by Daniel P. Moynihan (1970) pp. 8–9.

The Brookings Institution—for passages from *Problems in Public Expenditure Analysis* by Samuel B. Chase (1969) pp. 2–3.

Dover Publications—for passages from *Social Thought from Lore to Science* by Howard Becker and Harry Elmer Barnes (1961) pp. 680–681.

Duell, Sloan, and Pearce—for passages from *Frank Lloyd Wright on Architecture* by Frederick Gutheim (1941) pp. 147–148.

*Economic Journal*—for passages by A. R. Prest and Ralph Turvey, No. 75, December 1965, pp. 718–719.

Faber and Faber—for passages from *Towards an Organic Architecture* by Bruno Zevi (1950) pp. 67–70.

Free Press—for passages from *The Active Society* by Amitai Etzioni (1968) p. 283.

*General Systems Yearbook*—for passages from vol. 9 by O. R. Young (1964) p. 61; vol. 10 by Herbert A. Simon (1965) p. 76.

*Harper's Magazine*—for passages by Paul Goodman, January 1964, pp. 74–75.

Harvard University Press—for passages from *The Intellectual Versus the City* by Morton and Lucia White (1962) pp. 162, 235.

H.M.S.O., Civil Service Department, London—for passages from *Current Issues in Cost-Benefit Analysis* by H. G. Walsh and Alan Williams (1969) p. 6.

Johns Hopkins Press—for passages from *Operations Research and Systems Engineering* by Ralph E. Gibson, edited by Charles D. Flagle (1960) pp. 58–59.

International City Management Association—for passages from *Policy Analysis in Local Government* by Kenneth L. Kraemer (1973).

*Journal of The American Institute of Planners*—for passages by Gilbert Herbert, 29(3), p. 198; by Morris Hill, 34(1), January 1968, p. 21; by George Noel Kurilko, 34(3), May 1968, p. 198; by Michael B. Tietz, 34(5), September 1968, p. 307.

McGraw-Hill—for passages from *Architectural Record* 154(2), August 1973, pp. 65–66; from *House and Home* 44(5), November 1973, pp. 65–66.

Oxford University Press—for passages from *Oxford Universal Dictionary* 3rd edition (1955) p. 2115; from *The Political Philosophies of Plato and Hegel* by M. B. Foster (1968) p. 189.

Prentice-Hall—for passages from *Modern Political Analysis* by Robert A. Dahl (1963) pp. 9–10; from *The New Utopians: A Study of Systems Design and Social Change* by Robert Boguslow (1965) p. 66.

Scribner's—for passages from *The Dictionary of the History of Ideas* by G. N. Giordano Orsini, edited by Philip P. Wiener (1973) vol. 3, p. 421.

*Town Planning Review*—for passages by Nathaniel Lichfield, number 35 July 1964, pp. 160–169.

U.S. Agency for International Development and Department of Housing and Urban Development—for passages from *Industrialized Housing* by Ian Donald Turner and John F. C. Turner (1972) pp. I, 5–6.

U.S. Government Printing Office—for passages from Gunnar Myrdal quoted in *Industrialized Housing* (1969) p. 167; *System Program Management Procedures* by U.S. Air Force Systems Command, (1966) 375–4 pp. 1–69; 375–5 pp. 1–88.

University of California Press—for passages from *Systems Analysis as a Technique for Solving Social Problems—A Realistic Overview* by Ida R. Hoos (1968) p. 56.

The University of Chicago Press—for passages from *The State of the Social Sciences* by James G. Miller, edited by Leonard D. White (1956); from Paul Diesing in *Ethics* 65(1), 1954, p. 18.

University of Michigan Press—for passages from *Two Person Game Theory* by Anatol Rapoport (1956) pp. 5–6.

*Urban Land*—for passages by Alberto F. Trevino, Jr. in July/August 1970, p. 4.

John Wiley—for passages from *Introduction to Engineering and Engineering Design* by Edward V. Krick (1965) pp. 211–212; from *Modern Organizational Theory* by Chris Argyris, edited by Mason Haire (1959) pp. 124–125.

# Contents

# General Editor's Preface

Since time immemorial planners and architects have tried to systematize their methods of designing and carrying out the forms of buildings and of towns—to be philosopher-designers competent to formulate, without error, schemas for any predetermined purpose.

As our author points out, the strictures on building and planning in older times were minimal, amounting to little more than the well-known properties of materials and structural devices and the whims of priests and kings. As a result, analysis was generally identified with *intuition*, and efficiency of result, with norms of aesthetic *beauty*. Abstract, ideal configurations flourished—models of design that seldom underwent "testing" in the modern sense. When practical efficiency alone was sought, as in colonial towns or military camps, the layout was generally left to the equivalent of our engineers, who were usually of a military extraction. Examples of this were the Roman *castrum* and the medieval *bastide,* both of which were often planted in disputed frontiers between rival nations.

Today the designer of buildings and towns and cities has to answer to so many interested parties—multiple patrons, competing users, and unpredictable future residents—that there has arisen a strong movement to develop *controlled* methods of projecting and carrying out design procedures so as to arrive at a perfect synthesis by a reasonable, economical path. A considerable number of architects and planners are beginning to use what is in a general way called "the systems idea" by the author of this book. Some of the most promising techniques have been developed, in an historical irony, by the contemporary military-weapons industry. The wheel turns!

In our cities and planning series, this volume is one in which we are attempting to explore the historical precedents and future potentialities of some of today's more compelling ideals and procedures of design. The series-editor is preparing a volume of this type to deal with the purely visionary planning of our century.

G.R.C.

# I
# The
# Systems
# Idea

## 1. Introduction

Architecture has traditionally been produced by the interaction of a designer's experience, intellect, aesthetic sensitivity, and common sense. This interaction usually occurred within the established context of a style and of the client's mandate—with the consequent imposition of both objectives and constraints. A city, when it was consciously planned or designed at all, drew largely upon the same creative sources, with correspondingly larger but more poorly defined sets of objectives and constraints. Within the last decade, questions have arisen as to the adequacy of this historic mode of form-giving for both cities and buildings.

Critics of the traditional mode have argued for a greater systemization of the planning and design process; their premise is that things have simply become too complex for traditional methods.[1] Preindustrial cities and architecture were relatively simply artifacts. For one thing, construction techniques were largely determined by the materials employed and, given the relatively narrow choice of available material within a particular culture, techniques remained traditional. Intuition, in the context of this tradition, seemed a sufficient guide for innovative elaboration upon established methods. Thus, Eliel Saarinen could say that the medieval master builder was an intuitive genius who simply knew—he felt in his bones—how to put the stones together to create his architecture and, by extension, to build his cities.[2] Others, viewing the problem in varying contexts, arrived at essentially similar conclusions.[3]

The modern movement in architecture questioned this tradition. While it can be argued that the modern movement has only occasionally produced architecture or city design at the level espoused in its rhetoric, there was a break in the facile continuum of stylistic evolution. It can also be argued that, relative to preindustrial societies, contemporary architectural problems are of a wholly different order of complexity; there is, in fact, significant change in both the scale and scope of projects.[4] Today's large-scale architectural project encompasses more of an economic, social, and political nature than was encountered in the simple building of earlier physical structures. This added complexity demands a coherent ordering or organizing of the planning and design effort; in short, it requires a consciously articulated approach to the problem-solving process.[5]

In a historical context one can observe that, while the modern movement's grand transition in architectural form and style originated at the turn of the century if not earlier, it is only now that there has been a revolution in analyzing the thought processes encompassed by planning and design.[6] There was a good deal of rhetoric by the Futurists, Le Corbusier and others about the industrialization of architecture, yet only now are we experiencing "industrialization" in thinking about how one gives form to architecture and cities. This revolution in thought has generated a number of "approaches" to planning and design problem-solving—an approach being simply the conceptual framework within which one can engage a particular set of phenomena. Presumably some approaches are better than others in the sense that they are more efficient or effective in the solution of a specific problem. The demand for an "approach" is then a recognition that ours is a world far removed from that of Saarinen's medieval master builder: we now need to be self-conscious about how we proceed to solve the problems of planning and design.

While interest in *systems* is stimulated by the obvious need and desire to improve design and planning decisions, a much neglected impetus for systematic thinking has been the requirement to justify decisions once they have been made. In traditional architecture one was usually dealing with a single client and in many instances an autocratic one. In that context, decisions needed no justification. For example, if Louis XIV was shown a drawing by Le Nôtre, the issue was simply whether he liked it or not. The king's fancy was sufficient justification for proceeding with the project. Questions of design process, cost-benefit analysis, rationality of the solution, and so forth would have been both irrelevant and tedious. Such is not the case today when the client is very likely to be a corporate body, a community, a building committee, or some other potentially divisive group having diverse objectives and interests in the particular urban or architectural problem. Such a condition establishes a pressing need for explanation and justification to the various participants in the design process; assessment techniques and explanations of the "rationality" involved are very much in order. This is certainly evident in that great exemplar of the systems approach —a national defense department—where justification of a solution is

2

at least as much a part of problem definition as is the planning and design itself.[7]

Of the approaches available today, systems analysis is one of the more used and abused. Rationalizations of the planning and design processes have generally consisted of systemization of procedure, and the most fashionable methodology is systems analysis itself. This has resulted in a great deal of verbiage and some occasional nonsense but also in the generation of some fruitful thought about problem-solving in architecture and urban planning. Use of the systems idea in these areas is a natural extension of its application to other problems; supposedly systems analysis has been instrumental of late in the production of just about everything from children's toys to moon rockets. While one tends to be skeptical about the utility of this very chic analysis when it is applied to social, economic, political, and even psychological matters its manifest usefulness in space and military systems design is quite impressive, and the question arises: Is the transfer of the techniques to other types of problems possible; and if so, useful?

This question is partially answered by the many successful applications of the systems approach in situations where complexity must be arrayed within a coherent intellectual framework and where existing models or theories of how to proceed have proved inadequate in some way. This in part accounts for the appeal of systems analysis to the architect and planner; both are now faced with problems of increasing complexity, and the traditional mode of problem-solving is often found wanting.

It must be admitted from the outset that this trend towards rationalization of technique and methodology in architecture and planning is not without its critics. The idea of a systematic approach to architectural planning problems is seen by some as an inhibition; a tacit acknowledgment of lack of imagination or, at least, an unnecessary suppression of one's intuition. In the extreme, such systemization is denounced as a crutch for the untalented since there is no "art" in it. Critics of this persuasion seem to think that one's art, one's personal insights and perceptions, one's "feel for the situation" is sufficient guide to the problem-solving process, if indeed, planning and design are recognized as problem-solving processes at all. Why introduce, they ask, the needless jargon of the analyst accompanied by his unintelligible diagrams and interminable computer print-outs; is this not simply a palpable manifestation of the latest architectural fashion?[8] Other critics, less convincingly, have seen in the systems approach a tendency toward political deviousness in solving social problems.[9] They argue that it is a technocratic diversion from a more fundamental and searching analysis of society's problems. Finally, there are those who see the approach as an attempt to institutionalize simplemindedness in urban and architectural problem-solving.

The purpose of this book is to introduce the systems idea as objectively as possible. Given the brevity of the text, however, there are two necessary constraints: first, only those aspects of the systems idea which are of immediate consequence to the architect and urban planner are discussed. A second constraint involves our definition of urban planner and architect; they are viewed in a somewhat traditional sense as those interested in the city as a physical system or artifact. They do, of course, perceive the city as a container of social, economic, and political phenomena, and they appreciate the significance of these activities in giving physical form to the city, but their primary concern is for the artifact itself—the process of giving form to this artifact; in short, as planners and designers of physical systems. This is perhaps just another variant expression of the traditional architectural concern with "form following function." In this sense, social, economic, and political phenomena are here viewed as the determinants of form: they are functions to which the planner and designer must fit the forms of the city. And, while examples of the systems approach will be provided both for the analysis of social activity and the determination of physical form, it is the latter upon which this effort is focused.

The structure of our presentation is simple: first to define the idea in terms of its essential theoretical content, then to illustrate its actual use in architectural and planning practice and, finally, to assess its probable utility to both architect and planner. In order to exemplify uses of the systems idea diagrams and illustrations are taken directly from current works wherever possible. These, along with substantial direct quotation, will give the reader an indication of the wide diversity that exists in both the perception and use of the systems idea; it should also help correct any unwarranted syncretism on my part. Because of its brevity, the text is complemented by a selective bibliography and is amplified in extensive notes.

## 2. The Systems Idea in Outline

The systems approach as an operable concept is defined throughout this book in various ways: by formal definition, by tracing its evolution, and by describing practical applications of it in the real world. Our purpose in this chapter is to delineate the essential content of the idea.

Initial questions arise as to what an "approach" is. Quite simply it is a thought-model; a guide to attacking a problem. Presumably, some approaches are more fruitful in problem-solving than others. The essence of approaches is caught in Henri Poincaré's statement that bricks are to a house as facts are to a science. Just as in building a house, a blueprint is used to bring about the house's realization—an instrument to bring order to the bricks—so the thinker can use a theory or model to bring order to his facts. Both house and thought employ a patterning device which will impose a coherence: a blueprint of relatedness, a "system" which will integrate the elements and make of them a coherent whole.[1] Karl Mannheim says that the question is even more fundamental than this. He asserts that in all problem-solving there must be an image of procedure—of process—in the mind of the problem-solver. It is unavoidable; whether latent or manifest, implicit or explicit, one necessarily generates an image of how to proceed. Mannheim further asserts that the model developed by a problem-solver has ideological implications: it is the product of one's experience. This latter point will be significant in assessing the usefulness of a systems approach in architectural and urban problem-solving.[2]

The systems approach means many things to many people: a philosophy or world view,[3] a robust methodology for solving diverse problems, a formalized version of common sense, or institutionalized simplemindedness. The reasons for this diversity of assessment will become apparent later. Since our objective here is explanatory, we can focus upon two significant aspects that are both pervasive and essential in all definitions of the systems idea: *holism* in viewing the reality perceived, and *rationality* in attempting to deal with the problems that exist in that reality.

**Holism.** Holism refers to a perception of the relatedness of things in approaching reality or a problem. One abstracts from reality and forms an image of the interdependencies which exist among diverse elements. If one can conceptualize a useful network of these interdependencies, one has in fact defined a system. Thus, the systems analyst views things comprehensively in an attempt to perceive all the elements and relationships which might exist in a problem situation; he perceives a whole: a system. In a way, this is only common sense: trying to see how all elements of a problem relate to each other and with what significance. The systems approach is a way of dealing with complexity in the analysis of the organization of things as distinct from a fragmented or piecemeal approach.[4] As Chris Argyris has pointed out, there is a basic structure to the organization of reality which transcends usual depictions of it; it has systemic properties. Based upon this, Argyris has evolved a conception of organizations as systems which closely coincides with that expressed by systems engineers in viewing "hard" systems:

> An organization is characterized by an *arrangement* of parts that form a unity or whole which feeds back to help maintain the parts (Wiener). A "part" of an organization is actually an "organic" part in that it exists by virtue of its position in the pattern that is the whole (Kluckhohn). The whole, in turn, may be differentiated from the parts along two dimensions. First, the whole has a different boundary than any given part (or subset of parts) (Herbst, Simon).

Second, the functional unity of the whole displays properties only revealed in the actual process of full operation of the whole (Kurtz).

These propositions have led the writer [Argyris] to form his own (very) tentative conceptual definition of organization. An organization is

1. a plurality of parts
2. each achieving specific objective(s) and
3. maintaining themselves through their interrelatedness and
4. simultaneously adapting to the external environment, thereby
5. maintaining the interrelated state of the parts.[5]

Clearly, a significant element of this holistic or systems vision is the relationship of part to whole. This is expressed strongly in the idea of "suboptimality" as developed by economists and systems engineers; an idea which has significant utility in both architecture and planning as it focuses on relationships with which both inevitably must grapple. Suboptimization is the attempt to optimize the performance of a particular subsystem without regard to the effects of this optimizing action upon the larger whole of which the sub-system is but a constituent part. These effects are sometimes known as "spillover effects": the unintended consequences for other constituent subsystems generated by the sub-optimizing action.[6] Such effects ultimately can be positive or negative with respect to overall system performance, but it is the latter which obviously presents the problems. Suboptimization occurs when a decision-maker seeks multiple goals involving multiple subsystems as opposed to a single objective. In short, it is endemic to urban problem-solving.

An interesting example of suboptimization in urban problem-solving is provided in the history of transportation planning in New York City. There, transportation planners have historically viewed their decision-making role as one primarily directed at the ordering and ultimate optimizing of New York's transportation system. In accomplishing this, the planners have implicitly, and sometimes explicitly, discounted the effect of transportation subsystem optimization on other systems of the city, specifically those of a social and economic nature.[7] By definition they became suboptimizers. This perception of role, for example, resulted in advocacy of the often suggested, often rejected, Lower Manhattan Expressway. In pushing for this project, the advocates simply discounted the "spillover effects" of their optimizing activity upon other subsystems, for example, the Jewish, Italian, and Chinese communities in the Lower East Side and the small merchants along Canal Street. In addition to community fragmentation, many businesses would have been driven out of the area had the Expressway gone through. The systems analyst would conclude that the planners, in seeking optimal conditions for a particular subsystem (transportation), created negative spillover effects on the social and economic subsystems because of an implicit discounting of consequences in the decision-making process.

We should, however, be cautious in our condemnation of this transportation planning model because it is essentially that of many city planners. Each urban decision-maker is eager to provide optimal conditions within his own system. Each tends to neglect the overall effects of his decisions upon other systems within the urban environment. This condition has been abhorred by many thinkers in planning and is the basic reason for turning to a "systems" or, as it was traditionally known, "comprehensive" view of the city. By way of illustration, Earl Levin draws an analogy between planning practice and political Balkanization:

> By Balkanism, I mean the failure to see the whole city and the failure to see beyond the present movement; the habit of looking at everything in the city as though it were not a part of the city at all, but rather an isolated event, or even perhaps an event occurring mainly as part of one's own private and immediate world. The speculative builder sees his apartment block simply as profit on his own investment; the corporation sees its office building as the center of its livelihood; the traffic engineer sees the city in terms of traffic movements and the desired lines; the city engineer sees it in terms of sewer and water services; and even the city council sees the city mainly in terms of tax revenue mil rates. Almost no one sees it as an organic whole, with each part functionally dependent upon the other; almost no one thinks about the changes in the nature and function of the city in the future—even in the foreseeable future—which result from their actions in the present. . . . Planning is not city beautification; nor is it the construction of projects such as housing or highways; nor is it the application of bylaws and regulations for the control of development. It is not merely any of these things, and it is something more than the sum of all of them. It is a living process: the process whereby the wishes and preferences of . . . [those] who make up the community—become transmuted into the physical form of the city.[8]

It is this concern with a part at the expense of the whole which has resulted in so many "solutions" to urban problems that actually contribute to our present overwhelming problem. The obvious need for a holistic view to counteract these fragmented and partial efforts is a significant impetus behind "systems approaches" to the planning process.

**Rationality.** The systems analyst is not only holistic in vision, he is also rational in his method and procedure of problem-solving. The expression of this rationality is probably the most readily identifiable characteristic of the systems approach. Depending upon the relative emphasis of the holistic or the rational content of the idea, the approach itself is variously called rationalistic [9] (or simply rational and comprehensive); [10] "systematic" or "praxiological" (i.e., from the science of efficient action); [11] rational choice; [12] "rational-deductive" or "synoptic ideal." [13] In urban-planning thought, this systems approach has grown out of what is traditionally known as "comprehensive planning." [14] In most expressions of the model the idea of rationality is prominent and, consequently, certain specific techniques are immediately identified with the approach: simulation and modeling, probability and statistics, optimization studies, reliability studies, economic analysis, and project-management techniques. The purpose of our book is to provide an understanding of the conceptual framework or "rationality" which generates these techniques rather than to explore any of these particular techniques in detail.[15]

Fundamentally, being "rational" has meant having a problem-solving orientation.[16] In contemporary philosophy, the idea would appear to originate with John Dewey, who saw rationality as problem-solving: "the problem fixes the end of thought and the end controls the process of thinking." [17]

Such a problem-solving process is based upon accepted deductive scientific method. Essentially it involves:

1. the observation of phenomena.
2. the development of a hypothesis to explain what one has observed.
3. experimentation to test the hypothesis.
4. development of a theory.
5. management of the environment through the use of the theory.[18]

While this scientific method is the theoretical foundation upon which the systems-analytic approach has been formulated, in practice it is largely the working out of an economic rationality: one seeks the most efficient solution to a problem through the establishment of an optimal relationship between the resources expended and the results obtained; fundamentally a process of assessment and choice. This has led some to consider such assessment procedures as cost-benefit analysis, cost-effectiveness analysis, and cost-revenue anaylsis to be the systems approach itself. While this is a bit extreme, the fathers of the systems approach in the U.S. Defense Department, Charles Hitch and Roland McKean, specifically identified it with economizing.[19]

More will be said about these assessment techniques later, and while it must be admitted that there is no clear consensus definition today of just what the systems approach is,[20] the prevailing view is to consider it a union between the old idea of holism (or organicism) with newer (essentially post-World War II) analytic techniques of operations research. In urban problem-solving, the emphasis to date has been upon models of economic choice. The architect and planner, in attempting to define the particular configurations of their systems, have looked systematically (or organically) at the internal composition of the system (endogenous variables) and at the external relationships which the system exhibits (exogenous variables) and have searched for that structure of relationships which would ultimately yield the highest ratio of benefits (or objectives and values) achieved in relation to the costs of their achievement.

A description of this approach is provided by Charles Hitch and Roland McKean:

> It cannot be stated too frequently or emphasized enough that . . . [systems analysis] is *a way of looking at problems* and does not necessarily depend upon the use of any analytic aids or computational devices. Some analytic aids (mathematical models) and computing machinery are quite likely to be useful in analyzing complex problems, but there are many . . . problems in which they have not proved particularly useful where, nevertheless, it is rewarding to array the alternatives and think through their implications in terms of objectives and costs. Where mathematical models and computations are useful, they are in no sense alternatives to or rivals of good intuitive judgment; they supplement and complement it. Judgment is always of critical importance in designing the analysis, choosing the alternatives to be compared, and selecting the criterion. Except where there is a completely satisfactory one-dimensional measurable objective (a rare circumstance), judgment must supplement the quantitative analysis before a choice can be recommended.[21]

But Herbert Simon cautions us on the present state of the systems art:

> In both science and engineering, the study of systems is an increasingly popular activity. Its popularity is more a response to a pressing need for synthesizing and analysing complexity than it is to any large development of a body of knowledge in a technique for dealing with complexity. If this popularity is to be more than a fad, necessity will have to mother invention into providing substance to go with the name.[22]

## 3. Holism: Conceptions of System

In the Oxford dictionary, a system is defined variously as:

- An organized or connected group of objects.
- A set or assemblage of things connected, associated or interdependent, so as to form a complete unity; a whole composed of parts in orderly arrangement according to some scheme or plan; rarely applied to a simple or small assemblage of things (nearly = "group" or "set"). The whole scheme of created things; the universe.

- Physics: A group of bodies moving about one another in space under some particular dynamical law, as the law of gravitation.
- Biology: A set of organs or parts in an animal body of the same or similar structure, or subserving the same function; or the system; the animal body as an organized whole; the organism in relation to its vital processes or functions.
- Scientific: A group, set or aggregate of things, natural or artificial, forming a connected or complex whole.[1]

To introduce a bit of dictionary precision to the methodologies of architecture and urban planning, it may be useful to distinguish between simple and complex systems if only because there are many definitions of the former and much conjecture about the latter. For example, if we choose to view a pencil as a system, it is a relatively simple matter to see how one element—graphite—is secured to another element—wood—and capped by a third element—eraser—to form a system. It is also easy to define the system's normal function as an instrument to mark and erase lines. But if the system of concern were a city, could there possibly be such simple perception of the system's elements or function?

**Hard Systems.** In applications of the systems approach, a clarity of vision exists only in the "hard" systems area, roughly defined as systems engineering where the chief concern is with a physical system and the significant variables are normally quantifiable. Such systems have provided much of the contemporary format for the general development of the systems idea. In that field, use of the concept is consistent with our earlier dictionary definition of a system. Typical definitions employed include: "a whole composed of interrelated parts," [2] "a set of objects together with relationships between the objects and between their attributes," [3] "a set of elements standing in interaction," [4] or "a whole which is compounded of many parts. An ensemble of attributes." [5]

Systems engineers have also devised a useful format in exploring the idea of the "black box." In this view, the emphasis is upon what goes into and what comes out of any given system: input and output. The system itself is thought of merely as a transformational device or black box. About the contents of this box, we know virtually nothing except that it contains a number of elements in interaction—a system. When we provide an input to the black box, a transformation occurs within the system which results in an output (Fig. 1). This leads to a slight variant in the definition of a system:

Any process that produces a response (output) when an excitation (input) is applied to it, can be called a system (the term transmittance is also used).[6]

We can relate this to our above definitions by adding the essential "interrelated parts." An example of this is the business firm.

Such an enterprise may be viewed as a system with inputs and outputs of materials, money, information, etc. But, then, this enterprise is composed of a number of major subdivisions, each directed primarily to a specific type of activity, for example, sales, purchasing, and accounting. These major subdivisions of the system are called subsystems. The subsystems are closely interrelated parts of the whole; much information flows between them, and their actions are interdependent like organs of the human body. In many instances the output of one subsystem is the input to another.

Of course, these subsystems can be further subdivided. If we look inside each of the "department black boxes" we will find additional subsystems. The accounting department, for example, contains a billing subsystem, a payroll subsystem, an accounts payable subsystem, and others. This process of subdividing can be continued until further subdivisions become irrelevant to the problem at hand. For example, although a typewriter is a collection of many interrelated parts, as a part of the purchasing subsystem it may be only the machine as an entity that is of interest. Any system subdivision whose constituent parts are not a matter of concern in the problem at hand is ordinarily referred to as a component of that system. Note that what is considered the system in this example, a manufacturing firm, in some other instances would be a component of a much larger system. The economist in viewing our national economic system treats the individual firm simply as a component. In this case the particular make-up of the firm is of no interest; individual firms become "black boxes," components of the system that he is considering [7] (Fig. 2).

To relate the economists' concerns to those of the architect and planner, we might look at housing. If the problem is to design a single house, for example, the house may be considered as the system. The same house, however, would be a subsystem if the problem were to plan a neighborhood. Similarly, it could be viewed as a component element of a subsystem if the problem were to plan a new town comprising several neighborhoods. The definition of system is clearly a function of one's objectives in viewing a particular problem situation.

**Soft Systems.** The systems approach was late in developing in the social and behavioral sciences, where definitions of systems and relationships are not as easily arrived at as in physical systems and design analysis. Consequently, definitions of system in these areas are less clear and perhaps inevitably less satisfying.[8] For example, Robert A. Dahl, a political scientist, says: "Any collection of real objects that interact in some way can be considered a system: a galaxy, a football team, a legislature, a political party." [9] Dahl's use of the word "real" in this definition is troubling since most systems theorists admit of conceptual systems such as systems of thought or philosophy and "real" implies a physical entity. In most of the "soft" literature, a system is viewed as either real (i.e., physical) or conceptual; it can be a city (a sociophysical entity composed of many elements in interaction), or it can be a conception of interrelated urban processes in the mind of a planner. A system can also be simple or complex, large or small: an atom is a system, as is a building, a city, or a society. The complexity of social systems is reflected in Grinker's definition: "A 'system' is considered to be some whole form in structure or operation, concepts or functions, composed of united and integrated parts. As such, it has an extent in time and space, and boundaries." [10]

The best known systems theorist in sociology today is Talcott Parsons. To him, a system is defined by the interaction of two or more persons in which each actor attempts to account for the action of other actors or in which there are common goals of interaction. Dahl has expressed this idea of the social system in diagrammatic form. Of it, he says:

> According to Parsons' usage, a political system or an economic system would be parts, aspects or subsystems of a "social" system. This way of looking at the matter is illustrated (Fig. 3) . . . where AC represents the set of all political subsystems, and ABC represents subsystems that can be considered as either political or economic, depending on which aspect we are concerned with. Examples of ABC would be General Motors, the United States Bureau of the Budget or the Federal Reserve Board. [11]

Parsons's account is broad and abstract, but occasionally obscure. More useful and specific are constructs which attempt to account for interaction as an "exchange" of some sort. James Miller is a behavioral scientist who has developed such a thesis. He says:

> Systems are bounded regions in space-time, involving energy interchange among their parts, which are associated in functional relationships and with their environments. General systems theory is a series of related definitions, assumptions, and postulates about all levels of systems from atomic particles through atoms, molecules, crystals, viruses, cells, organs, individuals, small groups, societies, planets, solar systems, and galaxies. General behavior systems theory is a subcategory of such theory dealing with living systems, extending roughly from viruses through societies. . . . All behavior can be conceived of as energy exchange within an open system or from one system to another . . . and that all living systems tend . . . to maintain steady states of many variables . . . in an orderly balance but within a certain range of stability. [12]

One of the more elaborate uses of such a construct in viewing a community was made in the study of Stirling County, Nova Scotia. [13] In it, a community was viewed as a living, dynamic system which maintained an energy exchange among its various components. The totality formed a holistic community which tended to maintain a dynamic equilibrium.[14] The components of the system were perceived as individual human beings and, while each individual is physically detached from the system, he or she is nonetheless an integral part of the energy exchange continually in progress.[15] He is part of the organismic unity of the community. In it, energy is withdrawn from nature, then distributed through the system and component subsystems including themselves, and finally returned to nature. This essentially metabolic pattern is the basis of community integration:

> Moreover, although the whole is under the influence of its components, it has qualities and characteristics of wholeness which are the synthesis rather than the mere aggregation of parts.[16]

As an example of this synergistic or holistic quality of whole and part, Leighton cites government. He says that while all individuals contribute to it in various forms (taxes, voting, etc.), it has a quality of wholeness (as governor of the system) which is more than the sum of its individual governing acts; it is a "pattern composed of the total of these interdependencies . . . [a] whole that is more than the sum of its parts." [17] These interdependencies become more explicit when one attempts to focus on specific variables and trace their distribution or interactions throughout the system. An example of this is provided in the graphic representation of the interdependencies of an urban economy as reflected in the money flows within a city as illustrated in Figure 4.

One must recognize the dangers of the above way of looking at things; of abstracting from reality. Most prominent among them is what Whitehead called the error of "misplaced concreteness"—of confusing the abstraction one makes with the reality one perceives. To Dahl, for instance,

> To call something a system is an abstract way (or as some scholars say, an analytic way) of looking at concrete things. One should therefore be careful not to confuse the concrete thing with the analytic "system." A "system" is merely an aspect of things abstracted from reality for purposes of analysis, like the circulatory system of a mammal or the personality system of a human being.[18]

In conclusion, then, the most significant element in defining the system idea is its holistic content—an emphasis clearly expressed in Herbert Simon's definition of a complex system as

> one made up of a large number of parts that interact in a non-simple way. In such systems, the whole is more than the sum of the parts, not in an ultimate metaphysical sense, but in the important pragmatic sense that, given the properties of the parts and the laws

of their interaction, it is not a trivial matter to infer the properties of the whole. In the face of complexity and in principle, a reductionist may be at the same time a pragmatic wholist.[19]

The idea of a system is thus both simple and complex; in the abstract it is simple in its fundamental definition of any whole as composed of interrelated parts. This is expressed in the black box conception with its interrelated inputs and outputs. But in specific applications of the idea there is enormous complexity, especially in those dealing with social systems since boundary definitions are in such cases often nebulous: What is system as opposed to subsystem or element? What are the interrelationships that exist among them? This complexity is reflected in Kenneth Kraemer's succinct summary of the essential characteristics of those systems that concern planners:

1. The system constitutes a whole set of related things or events;
2. The whole is seeking to fulfill a set of goals;
3. The whole is composed of differentiable elements or subsystems, and the elements or subsystems are integrated in a patterned or structured form;
4. The elements or subsystems are in interaction, mutually affecting one another;
5. The whole system exists within an environment which is distinct and definable from the system itself;
6. A boundary differentiates the system from its environment;
7. The system is in constant interaction with its environment and producing outputs in exchange;
8. The system processes inputs into outputs through internal transactions in accordance with established needs;
9. The whole system seeks to maintain a state of dynamic equilibrium internally with its subsystems and externally with its environment. Feedback is the process by which the system maintains equilibrium and steers towards the system's objectives;
10. To maintain dynamic equilibrium, the system is in a state of constant flux or change;
11. The system has some mechanism for the control of its activities.[20]

In sum, the systems approach is an attitude of mind in facing complexity; it reflects a search for the interrelatedness of things in any problematic situation. As a planning tool, it means approaching the city as a very complex whole within which many elements act interdependently. The premise is that the city has a certain synergistic quality, that it is, in fact, more than the sum of its parts.

## 4. Rationality: Manifestations and Instruments

The systems analyst's reasoned approach to problem-solving is premised upon an objectivity in viewing problems. While there is no doubt that this is seriously attempted in most instances, there are substantial questions as to the possibility of objectively viewing social phenomena; such objectivity often appears to be a positivist illusion or a utopian hope. One's biography *is* reflected in one's constructs.[1]

Investigation of this premise also leads one to examine the possibility of collective rationality in social problem-solving. Much has been written on the potential of determining "rational" behavior in addressing issues which somehow involve a "public interest."[2] A particular rationality is emphasized—consciously or unconsciously—by decision-makers. Paul Diesing finds at least five basic models of rationality operable within any given society—technical, economic, social, legal, and political—which result in five distinct kinds of social decision-making.

For the systems analyst most significant are technical and economic rationality. Diesing, however, identifies technical rationality as the efficient achievement of a single goal. When there is a plurality of goals, their achievement is identified as economic rationality because of the trade-off considerations involved. These two types of rationality are usually the dominant forms within our society—sometimes to the point of near exclusion of all other forms. However, the criterion of efficiency

> applies only to means and not to ends or ultimate values, thus . . . the basic aims of life cannot be selected or evaluated by rational procedures: they must be dealt with by arbitrary preference or intuition, or by cultural and biological determinism.[3]

In developing this point, Diesing warns against the reduction of one type of decision-making to another. He is particularly critical of the present reliance upon technical and economic rationality as exclusive means of decision-making.

Diesing states that when one employs economic rationality the basic logic is one of allocation and exchange in a neutral environment.[4] In this process, the "ends are alternative when they have common means which are scarce, so that an increased achievement of one necessitates increased achievement of others . . . [as a result] economizing is an evaluation and selection of ends, and it occurs when two or more ends are in competition with each other."[5] For Diesing, *social* rationality is based upon interdependence and solidarity. Decisions so based will more fully integrate the society. *Legal* rationality concerns "an order of availability . . .": it determines what resources are available to each legal person, what actions each person must perform and, in turn, what expectations one might have of others. *Political* rationality improves the quality of all decision-making structures in a society.[6]

**Problem-Solving Orientation.** The network of activities which we usually associate with physical planning is one which requires the development of guidelines, criteria, and specifications for the creation and modification of physical environments,[7] coupled with the development of mechanisms useful in making the desired "planned" or "designed" environmental condition a reality. The initial step in this process normally occurs when an actor—political, institutional, corporate, or individual—decides to attempt the modification or creation of an environment. The actor generally asks an environmentalist—planner, architect, engineer—to establish the "plan" to be employed in this modification and to help in bringing it to fruition. This "plan" is, in fact, the specification of a desired future environment or, more generally, a desired future state of affairs. The act of developing these criteria and programs is variously called planning, architecture, environmental engineering, ergonomics, and so forth, depending both upon the scale at which the environment is to be modified and upon the specification by the initiating actor of those variables he considers subject to modification. In effect, this specification provides both a definition of the purpose and scope of the intervention by the environmentalist and the setting of constraints within which he must work.

Implicit in such a conception of planning are suggestions of a disciplinary "turf" which the environmentalist himself may wish to define tightly or loosely, depending upon his interests and capacities, his perception and understanding of the problem to be solved, and the institutional constraints within which it is to be solved. Environmental planning in this context has associations with many varied activities among which are analysis, scholarship, and creation in an artistic sense. However, generically the environmentalist is *solving a problem* that exists within the environment. This problem may be defined in terms of largely quantifiable variables as in the case of the planner or systems analyst, or it may be essentially qualitative (and specifically aesthetic) in the more traditional architectural sense.[8] In any case, the creation of an environment demands a problem-solving orientation by the environmentalist. Here, as with the systems idea itself, there are elements of a built-in bias; specifically of pragmatism and the reductionism exemplified by Dewey and systems-oriented thinkers generally.

A problem is generated by the desire to transform one state of affairs to another:

> There is an originating state of affairs referred to as *State A* or as the *Input;* similarly, there is a state of affairs (objective, result) which the problem-solver is seeking a means of achieving and which is called *State B* or the *Output.*[9]

The black box format to which we have referred is assumed in this definition, which can be illustrated by an urban slum area. We might assume that the slum exists in a dilapidated physical condition and is suspected of generating various individual and social pathologies (Input). The planner hopes to transform the area into one of safe and sound physical structures and to provide the amenities he imagines necessary for a wholesome social life (Output). To accomplish this, he might use any number of means to his end, e.g., rehabilitation, clearance and reconstruction, employment opportunities, various recreational facilities, and so forth, singly or in combination. These of course are the "alternative systems" or "options" of the systems analyst. Each alternative has certain advantages and disadvantages relative to other feasible alternatives. Assessing the cost-reward structure of the options is a major part of problem-solving; to some, it is the problem. As Krick says:

> If there are no alternative means of accomplishing the desired results, there is no problem. Similarly, if all possible solutions are equally desirable, no problem exists. A problem involves more than finding a preferred method of achieving the desired transformation. For instance, most persons are not indifferent to the different costs, speeds, degrees of safety, comfort and reliability associated with alternative means of travel.[10]

To find out which alternative is preferred, the planner refers to a criterion or some mix of criteria (a "criterion function") which are essentially operable measures of determining the achievement of previously formulated objectives. In effect, these criteria provide a checklist for the evaluation of various alternatives.

We must note also the conditions under which a planner "solves" problems. There are always various elements over which the planner has no control. These are variously called constraints, restrictions, or exogenous (i.e., external) variables. In the slum example, for instance, there may have been a maximum budgetary figure for the modification that was imposed on the planner by a higher authority or a specified maximum number of either the persons who could be relocated or the population densities that were to be realized in the final output state. In city planning, one is also dealing with a dynamic system: the city today is not what it was one year ago nor what it will be in another year. Conditions, constraints, and objectives are always in flux. This presents a major difficulty for the planner: how to perceive the problems—the urban system.

**Abstraction and Model-Building.** Seeing problem-solving as the transformation of system states and arriving at some operational definition of the systems transformed are relatively simple tasks. Once certain premises are established, the logic of procedure becomes self-evident. But a determination of the transformation itself revives many of the problems we encountered previously: the variable nature of urban decision processes, their contention with market and political processes, etc. In practice we cannot expect different actors to assert compatible definitions of what is to be done with the urban system, nor even to agree on what its problems might be.

A determination of the problems involves the way one abstracts from reality and reconstructs images or models of it. Such processes are basic to any rigorous approach to planning, be it a specifically systems model or not. Implicit in much of our previous discussion of suboptimality is the idea of an underlying variation in conceptions of the urban system that roughly correlates with variation in the discipline or field—the perspective—with which a decision-maker views the system. This perspective defines the way in which one perceives a problem. The issue here is essentially the requisite abstraction process one goes through in first defining a system and, secondly, the problems within it. The architect, the economist, the sociologist, each has his own conceptual model of the system and its problems.

Unfortunately, the word "model" is becoming as overused and misused as the term "system" itself. The two concepts are, however, inextricably interrelated. The concept of system cannot adequately be examined unless some investigation is first made of this question of abstraction and the resulting synthesis in model-building. To summarize: the system one constructs is formulated in terms of a model and, conversely, the model is usually that of a system.

In its simplest sense, a model is a *representation of reality*. Usually it is a representation of a system which is an intellectual construct of that reality. The model is arrived at through the process of abstracting from reality those aspects with which one is concerned. There is no "true" process to be emulated; even if one could somehow comprehend the totality of the situation, the resulting model would be so complex as to render it useless. This last point is significant: one builds models in order to solve problems, to understand or to control some aspect of reality. Some models are descriptive, others predictive or normative. One builds a descriptive model for purposes of communication and heuristics (discovery), and a predictive model for projecting and ultimately controlling the behavior of a system. A systems analyst has summed up this process in an uncharacteristically poetic fashion:

Man tries to make for himself in the fashion that suits him best a simplified and intelligible picture of the world. He then tries to substitute this cosmos of his own for the world of experience, and thus to overcome it. This is what the painter, the poet, the speculative philosopher, the writer, the natural scientist and the systems analyst do, each in his own fashion.[11]

Thus, the process of abstraction is complex, value-laden, and difficult. Protestations as to objectivity in the process are useless. Anatole Rapaport tells us that:

A key word . . . is *abstracted*. It implies that only the essential aspects of a situation are discussed in game theory rather than the entire situation with its peculiarities, ambiguities and subtleties. If, however, the game theoretician is asked, "What *are* the essential aspects of decisions in conflict situations?" his only honest answer can be "Those which I have abstracted." To claim more would be similar to maintaining that the essential aspect of all circular objects is their circularity. This may be so for the geometer but not for someone who distinguishes coins from buttons and phonograph records from camera apertures.

To be sure, the geometer deals not with "circular objects" but with circles. That is to say, the conceptual act of abstracting circularity from all circular objects was performed long enough ago to have been institutionalized in our language and in our science. Hence the geometer can assume that people who wish to study the geometric properties of circles will easily forget all the other properties of circular objects, such as color, the material from which they are made, or the uses to which they are put.[12]

The would-be model-maker of a social-activity system, or of our environmental (physical) system to contain it, is in a more difficult position than is the game theoretician. To appreciate his difficulty, we might begin with an illustration of the process (Fig. 5). We have the "real" world or that aspect of it with which we are concerned. We may wish to examine, discourse upon, manipulate or control this portion of reality. We first observe and measure certain aspects of the real world and then by correlation, correction, resynthesization, or the like—i.e. through iteration (repetition)—we build a useful model of the situation.

**Procedural Coherence.** The rationality inherent in the systems approach is illustrated by its focus upon procedural models; they, in turn, generate the flow and process diagrams and procedural charts that are prominent tools of the analyst. The concern of the analyst is to make the planning and design process rational so that it has coherence. He assumes that there is both a temporal and logical order to the events and activities we call design and planning and that there should, therefore, be a logical sequential order within which one engages them. John Dewey anticipated much of this thinking in his early assertion that a decision process has

five logically distinct steps: (i) a felt difficulty; (ii) its location and definition; (iii) a suggestion of possible solution; (iv) development by reasoning of the bearings of the suggestion; (v) further observation and experiment leading to acceptance or rejection; that is, the conclusion of a belief or disbelief.[13]

The following represents a recent contemporary consensual expression of this:

In the simplest form, the problem-solving steps which underlie planning are as follows: First, determine and define the problem; second, collect all relevant facts, consider all possible alternative solutions and choose a solution or solutions to solve the problem; and third, take action to solve the problem.[14]

To some, however, this is not a description of how we actually solve problems but a normative ideal of how they should be solved (Fig. 6). More particularly, it has been viewed as representing a scientific ideal gratuitously transferred to nonscientific concerns.[15]

The origins of the model are associated specifically with economic, mathematical, and statistical conceptions of rational behavior.[16] Certainly the basic "systems" expression of it is an economic model of decision-making and is identified as such by its authors.[17] A variation is provided in Irwin Bross's three-stage model:

(1) There are two or more alternative courses of action possible. . . . Only one of these lines of action can be taken. . . . (2 )the process of decision will select, from these alternative actions, a single course of action which will actually be carried out. . . . (3) the selection of a course of action is to be made so as to accomplish some designated purpose.[18]

The model is also basic to corporate planning. Koontz and O'Donnell say that such planners:

1. establish objectives
2. establish planning premises
3. search for alternative courses of action
4. evaluate alternative courses of action
5. select a course or courses of action
6. formulate necessary derivative plans [19]

There are nearly as many versions of these process models as there are people with faith in collective rationality, and the models vary in their emphasis, subtlety, and usefulness to the architect and urban planner. A consensus definition—really a digest of the assorted models —would approximate the following (Fig. 7):

1. define objectives and the terms of trade-off between them;
2. posit and examine alternative means of achieving the various objectives;
3. identify the probable consequences of each alternative means;
4. select that alternative with the preferred set of probable consequences.

When one looks for operational models of procedural coherence specifically directed at determining the configuration of physical systems—that which we traditionally know as planning and design—there is a plethora of speculative thinking but little of tested real-world utility (see Part II). One that has been tested is the U.S. Air Force Systems Command model which is delineated in a series of manuals on the systems-engineering process.[20] The systems of concern to the Air Force include ground-support systems such as air bases. This model is delineated here because it received some of the best systems-engineering thinking available during its development. It expresses in a coherent and manageable vocabulary a clear delineation of the planning and design process—something that could not be achieved in a brief digest of several models. The methodology is particularly helpful because it has been employed in the planning and design of very large and diverse systems involving the functional integration of numerous and complex subsystems.

Two general processes are developed. One concerns the *management* of a physical system from its inception to final use; the other deals with specifics of actual design or *engineering* of the system. Of the former, we are told:

A program for system management must recognize the natural order of actions during the life cycle of a system. The system must be first conceived; this is designated the "conceptual phase," during which the idea and requirement for a system are born. It must then be defined; this is designated the "definition phase," during which detail design, development production, and testing of the actual elements of the system must take place. It must then be delivered and put in use; this is the "operational phase," during which the system is ready to be operated and the mission of the system can be fulfilled.[21]

In practice, an additional "acquisition" phase is added so that actual systems development proceeds in the following sequence:

*Conceptual phase:* The objective during this initial stage is to define the basic requirements for a system and then to explore ideas for possible physical configurations which would support the requirements. This means definition of "integrated broad system requirements, a systems concept to satisfy the system requirements . . . and preliminary technical development plans." [22]

*Definition phase:* Here, the results of the conceptual phase are "translated into detailed systems and system element performance and design requirements." [23] This requires a definition of system elements in discrete and specific terms and a determination of both cost schedules and performance objectives before development of a system is attempted. During this phase, an overall systems specification is developed which incorporates both system performance and design requirements. The effort is essentially one of assessing whether or not effective use is being made of available resources in the development of a system. Also, there is a definition of intersystem and intrasystem interfaces in order to more precisely define design responsibilities for various "subsystem" planners. In net effect, the definition phase is one in which the system is given form and actually engineered: planned and designed.

*Acquisition phase:* During this phase, the system is constructed and acquired by the Systems Command. However, there is some system modification based upon redefinitions of system performance requirements as the process of acquisition goes on and more detailed design information is developed and fed back to the designers. This results in a continual updating of the detailed plan previously defined. There is also some testing of the performance characteristics in various areas to determine if, in fact, the system will meet the predetermined performance requirements. And there is also a verification of the validity of the previously determined specification-tree and a greater detailing of specifications where necessary. There is, in sum, "the accomplishment of preliminary and detailed design and performance of design reviews." [24]

*Operational phase:* This phase is self-explanatory; it consists of turning the system over to the user organization for final dispositon and a transfer to them of responsibility for maintenance of the system. Evaluation of the system and feedback into redesign processes still occur.

The second general process, that of the system design or engineering, is an aspect of the overall management process. Specifically, it is an attempt to define "a common system analysis process that leads to system definition in terms of performance requirements on a total system basis." As such, it is pragmatic holism: a response to "an emerging awareness of the need for and importance of *TOTAL SYSTEM DESIGN.*"[25] The engineering design of the system begins in the latter part of the conceptual phase and continues into the early operational phase. Specifically,

> the process is a method for defining the system on a total system basis so that the design will reflect requirements . . . in an integrated fashion. It provides the source requirement data for the development of specifications, test plans, and procedures; and the back-up data required to define, contract, design, develop, produce, install, check-out and test the system. . . . The two fundamental purposes of the system engineering management process are: (1) to establish a single analysis, definition, trade-off, and synthesis of requirements in design solutions on a total system basis and (2) to provide a clear and concise reference source for communication of selected system design solutions between [designers].[26]

To accomplish this, provision is made for a "detailed road map of engineering actions during a systems life cycle in their relative order of occurrence . . . [focusing on a concern for] deriving a coherent total system design to achieve stated requirements." This is a recurring emphasis and appreciation for the integral or total system orientation based on a recognition that choices in one aspect of the system will usually affect many other system variables.[27]

Operationally, the process is one which encompasses the definition of:

1. systems objectives
2. the "design to" requirements necessary to meet the objectives
3. the "build to" requirements which prescribe the ultimate configuration of the system to be delivered to the user
4. other requirements for procedural and logistic support.[28]

This is achieved by proceeding in a cyclic fashion through a sequence of design phases. This procedure is diagrammed in Figure 8. If employed with discretion, the process is one which will help to:

1. define the necessary elements that are required to fulfill the total system or project objectives;
2. develop performance, design and test requirements early as the basis of the integration and trade-off of system performance requirements, system elements and system and . . . design constraints;
3. provide the necessary criteria in the system performance/design requirements [for] general detailed specifications for evaluation . . . design and development and production effort against specified performance as the basis . . . for procurement;
4. define and control the intersystem and intrasystem interfaces at each step throughout the definition and acquisition process;
5. provide a framework of coherent system requirements to be used as performance, design and test criteria; serve as a course for development plans and so forth;
6. provide a functional system model for use in generating mathematical models including simulation techniques, to quantitatively evaluate system effectiveness before, during and after design, fabrication and test of the system.[29]

There is a strong parallel between this system-engineering process and the traditional design approach used in architecture since both are premised upon "form following function." [30] In Systems Command jargon, this means that when operational requirements are translated into a functional definition of elements,

a common reference point or functional basis is established for developing the elements of a system. This functional approach can assure that definition is on a total system basis in full recognition of all involved elements . . . [and] the application of the functional approach to types of systems results in the logical definition of systems elements. . . . [Therefore] it is necessary that all systems programs initially identify the primary mission functions and requirements and then identify the functions and requirements for supporting the mission as the only basis for defining total requirements and selecting systems elements. . . . These initial functions would then serve as a basis for developing more detailed operations functions in order to identify more detailed requirements in any needed trade-off studies.[31]

This breaking down of major system functions into an ever finer "functional" division of labor within the system is accomplished within the Systems Command's procedural network by the use of diagrams which are very similar to the "functional layouts" used by architects. The entire process should, therefore, be quite comfortable to the architect schooled in the "functional" tradition.

The precise configuration of the functional diagrams is determined by a process the Systems Command calls "indenture"; a process in which the broadest and most general functions (termed "gross functions") a system is to perform are continually subdivided into more specific and precise definitions of subfunctions.[32] For example, in the design of a house, the gross function might be defined as providing shelter for a family. This gross function would in turn be divided into so-called first-level functions, hypothetically, to provide a living room, kitchen, bedroom and bath. These first-level functions would then be further subdivided. For example, the kitchen function for a kitchen would be further subdivided into the provision of facilities for cooking, refrigeration, washing food, etc. As the sequential and increasingly precise definition of function occurs, there are a simultaneous series of trade-off studies amongst functions. For example, given limited resources, is it better to have a third bedroom in a house or 50 percent more living space in the living room? Which condition would adequately meet a client's objectives?

This process of functional indenture is illustrated in Figure 9 in which one of the first-level functions of a hypothetical system of facilities is broken down into subfunctions and, in turn, one of these subfunctions (i.e., the generation and distribution of power) is further subdivided into its subfunctions. The process, of course, would continue beyond this level. While architects have always asserted that form follows function, none have attempted to be this precise and specific in the definition of the functions to which their forms are fitted.

While we have with the AFSC model established a very fundamental conceptual similarity between the traditional architectural and physical planning process, interesting parallels in "form following function" obviously do not establish an identity. But that the parallels are more than superficial is revealed by comparing the AFSC model with more general models of the design process. A particularly good example of the latter is the design cycle conceived by Edward Krick (Fig. 10), in which the critical nature of the problem-formulation stage is well illustrated. It also demonstrates why design-process models generally are not immediately adaptable to a planning process directed at a determination of urban form: the "problem," the "functions," or the so-called "systems requirements" for urban systems still require definitions that are essentially political in nature.

**The Assessment of Alternatives.**    Most significantly, the rationality of the systems approach is expressed in various techniques of assessment and choice. Essentially all are concerned with determining the optimal choice when concerned with alternative courses of action. In practice this amounts to the working out of economic rationality and this gives the approach its usual identification with cost-benefit and cost-effectiveness analysis.[33]

The different techniques had early applications at the turn of the century in the economic analysis of railroads, and were institutionalized in The Flood Control Act of 1936 which provided a Congressional mandate to the U.S. Army Corps of Engineers stipulating that, in their projects, the estimated "benefits to whomsoever they may accrue" should exceed the estimated costs. Cost-benefit analysis has been used in the field of engineering economy for some time but only achieved wide recognition in government circles with the MacNamara changes in the Defense Department. It provides a numerical definition of the ratio of the benefits from a project to the costs of same after each is comparably discounted with respect to time at a minimum attractive rate of return (interest rate). Consequently, projects with benefit-cost ratios greater than one are economically feasible, those with ratios less than one are not. The technique is discussed and illustrated in Part II. It should be mentioned here that benefit-cost analysis has been subject to substantial criticism in that it deals only with quantifiable and, moreover, commensurable variables.[34] Specifically one must identify the cash flow of both benefits and costs on a project before the technique is applicable.

Another aid in assessment and choice is the trade-off study,[35] an idea not altogether alien to traditional architectural and planning thought. While it originates in an economizing approach to problem-solving, and is implicit in the previous discussion of technical and economizing rationalities as suboptimizing orientations, it is the same course architects follow when they decide to give up one point of view for another (oak paneling for plaster board).

In any real decision-making process, one is rarely pursuing a single end. There are usually multiple objectives and one must examine multiple, complex, and interrelated means to these ends. As a result, there must be some determination of the terms of trade-off between the several objectives: how much of one goal or value does one want as measured by the sacrifice of other goals and values that one is also seeking. Practically speaking, this involves the definitions of what combined set of objectives—what "objective function"—will be taken as the end sought. This is essential in any real situation as there are never sufficient resources to optimize all objectives.

The best examples are, again, provided by aerospace systems. In determining a systems program, for example, the various program managers must trade off between missile size (weight), rocket size (weight), control mechanisms, etc. Obviously, the engineer managing the whole system would prefer to have the maximum missile size, the maximum power plant capacity, the maximum amount of guidance, and so on. But all of these elements could not be packaged into a complete system in any reasonable manner because of limitations on weight, fueling, and the like: there must be trade-offs between systems and their component elements so as to determine the ultimate configuration.

City planning requires trading-off on a grand scale. A recent planning experience in the Bahama Islands provides a clear example.[36] Currently, the Bahaman government has a very limited budget. Due to an irresponsible attitude toward investment in social welfare on the part of previous governments many problems have arisen. These problems have generated urgent needs for the many physical support systems necessary to maintain activities in the islands. It is not unrealistic to assume that the entire national budget could be usefully absorbed in the redevelopment of support systems presently in decay—housing, transportation, communication, sanitation, etc. Alternatively, the budget could also be used in establishing new activities in health and education. This, of course, would lead to the consequent generation of new physical support systems. As a result of all this, the present governmental decision-maker must determine many trade-offs between various social objectives and their consequent physical support systems in order to build the social infrastructure necessary in the islands. Specifically, he must deal with such problems as the determination of how much of a scarce and common resource—money—should be devoted to health as opposed to education; or with respect to physical support systems, how many schools should there be versus how many clinics. Particularly significant in these determinations is the overriding question of whether the decision-maker will reflect a *total system* orientation or will simply choose to define the problem and consequently the "objective function" within the constraints of a narrow suboptimizing rationality, usually with respect to only one or a few activity systems, the latter attitude generating negative spillover effects upon other urban systems within the whole: the growth of highways at a cost to everything else is the classic example.

Trade-offs are a large part of the idea that Robert Venturi is attempting to get at in his "contradiction and complexity" in architecture. The client today is no single source but rather corporate bodies, building committees, community groups, planning boards, and the like. There is an inevitable diversity of objectives. On occasion, there may be a coalition of interests so that a coherent and unified design policy will emerge. More usually, the result is continuing tension and never a total resolution of conflicting aims by the various participants who are, in essence, directing the design or planning of their project. This builds a contradiction, a complexity, and an inevitable tension into the process for which the trade-off study might be a most useful planning tool. It would help to identify who is getting what out of the building process: what aesthetic, social, economic, political aims or goals are being satisfied, and at what costs in terms of other objectives and values.

## 5. Architectural and Planning Antecedents of the Systems Idea

In an age of rapid social and technological change, it is not surprising that we sometimes tend to regard the more significant contemporary ideas as being conceived yesterday, if not this morning. This is certainly true of that "new" idea: systems. It conjures up images of the space age, atomic submarines, and "rational" men in the Pentagon assessing the options.

Admittedly, it is the demonstrated utility of the systems idea in the fields of defense and space that has generated its appeal to others, including architects and planners. However, with respect to its holistic or essential systems content, the idea has been around long enough to have found expression in the thinking of Plato and Aristotle.[1] With respect to its rational content, it is more recent. While it may appear to be a latter-day expression of the Enlightenment enthusiasm for rationality in social affairs, it differs in that a limited perception of the rational—economic rationality—forms the foundation of today's systems studies, as we have seen.

Current use of the systems idea stems from several sources. Among the more notable were operational research studies in England during World War II. Led by Professor P. M. S. Blackett ("Blackett's Circus"), teams of scientists sought to improve military problem-solving with a special interest in antisubmarine tactics. There were similar operational research groups in the American defense establishment, leading ultimately to the establishment of RAND and other "think tank" organizations within which economists were prominent. Some of the economists who worked for RAND in the late forties later went to work for the Defense Department under Robert MacNamara in the early sixties where systems studies flourished.[2] In a broader context, there was the work, also begun in the early forties, of Professors Singer, Churchman, Cowan, and Ackoff at the University of Pennsylvania.[3] An increasingly significant stream of influence is General Systems Theory, which emerged partially out of organismic conceptions of biology applied to nonbiologic systems.[4]

It would be interesting to trace the development of both the holistic and rational content of the systems idea through history in general but our focus must be limited to those aspects which have found expression specifically in architectural and planning thought. Essentially, this means exploring only its holistic evolution. Holism has had great appeal to architects in particular as they have always acted as systems integrators attempting to tie together structural, spatial, and aesthetic systems within a unified or holistic entity—a work of architecture.

While holism has been pervasive throughout history, insight into its development is made difficult by the ambiguity of various words used to express the idea: system, function, organic, organismic, holism, and others.[5] The greatest ambiguity in expression has been generated by the use of various organic analogies; some regard them as metaphysical or romantic tendencies: the seeing of natural or biological characteristics in entities which are systemic in nature but which in fact may have no biological attributes at all. These biological overtones are reflected in the general intellectual climate of the time. Certainly the flourishing of biological sciences in the nineteenth century contributed not only to Social Darwinism but also to the conception of society as somehow analogous to biological organisms. Among sociologists, this position has now been discredited but occasionally the analogy is still encountered.[6]

Architects have also been prominent users of the natural or biological analogy; until recently, it was commonplace among them to include the word *organic* in laudatory descriptions of a design. Usually the alleged organic quality was left undefined but the mere use of the term identified one with a mystique of the anointed or, minimally, an association with Sullivan, Wright, et al. It was also something of a code word to identify those designs outside the International Style. Whatever its use, the word and the concept became firmly ensconced in architectural literature.[7]

The term made an easy transition from architects to urban planners. Some saw in an all too ill-defined organic approach to their work an invitation to biomorphism, thus providing a physical analogue to the excesses of Spencer and the organismic sociologists. It offered a nebulous mandate to curve streets in a spaghettilike fashion, weaving through the various "organs" of the city. It also furnished a ready-made vocabulary within which the planner could describe his ideas in such organismic terms as "circulation" with its chief organ the "heart" of the city presumably pumping the economic "life blood" of the community in order to nourish the neighborhood "cells" or "organic residential units." All this was controlled, of course, by a "nerve center" in the central business district.

Where did this organic thinking originate? Was it part of the continuum of holistic or systemic thought from Plato to Spencer? Perhaps, but it is difficult to be conclusive on this issue. There is, for example, organic, protosystems thinking in Vitruvius, and in Alberti the systems idea becomes quite explicit:

In order to therefore be as brief as possible I shall define beauty as to be a harmony of all the parts, in whatsoever subject it appears, fitted together with such proportion and connection, that nothing could be added, diminished or altered, but for the worse.

He even adds the seemingly inevitable biologic association by saying that this systemic quality "is but very rarely granted to anyone, or even to nature itself." [8] This association persists today within an esoteric specialty of the contemporary systems field known as bionics.[9]

Modern organic thinking in architecture seems to have had its genesis largely in Romanticism; the terms "romantic" and "organic" being virtually synonymous in architectural literature.[10] Romanticism in this context means the nineteenth-century reaction to rationalism: the reliance upon emotion and tradition in place of the earlier belief in intellect and thought.[11] Coleridge, Emerson, Thoreau, and Whitman were the principal romantic influences on the thinking of such pioneers of organic architecture as Furness, Greenough, Richardson, Sullivan, and Wright.[12]

The romantic reaction among other pioneers of the modern movement in architecture and planning (Ruskin, Morris, Howard) was essentially antiurban and principally so in reaction to industrialization.[13] Tonnies' earlier conception of organic or closely knit community (Gemeinschaft) found its way, via several organic sociologists, to Clarence Stein and Henry Wright and was ultimately realized in the planning of Radburn. We might also recall the praise that organically oriented critics such as Lewis Mumford have given to small, cellular communities, especially to Radburn and Howard's Garden City. In sum, we have the postulate by the romanticists of the small, cellular, cohesive, essentially medieval, urban form as the properly organic model for urban planning.

But there was more to these tendencies than a search for models of form. Robert Nisbet says that Ruskin and Morris, in reacting to industrialization, "called attention to the cultural and moral costs involved —the uprooting of family ties, the disintegration of villages, the displacement of craftsmen, and the atomization of ancient securities—the apostles of rationalism could reply that these were the inevitable costs of Progress." [14] Morton and Lucia White have related this to other aspects of social thought at the turn of the century, such as:

[William] James' idea that the big unit was hollow and brutal was applied with ease to the city itself by reformers like Jane Addams and John Dewey. . . . Progressive reformers continued to espouse the ideals of an earlier period in American history in their attack on big organizations. Jane Addams and Dewey could easily recall their own rural backgrounds, and Park could supplement his personal experience by reading German sociologists like Tonnies and American sociologists like Cooley, who glorified the face-to-face relationships of so-called primary groups like the family, the village, and the church in which people saw each other frequently and continuously over the years. . . . The basic strategy of this movement of urban thought was not to destroy the city but to recreate within it something like the spirit of life as it was lived in an earlier time.[15]

The pervasiveness of this type of thinking among architects, while emphasizing the general systemic aspects of the city—the interdependence of parts and so forth—also reflected a strong ideological bias in their romantic preference for the small community. This has some disconcerting elements. Walter Laquer points these out in an observation on the "voelkisch tradition" in German thought that significantly parallels some of the architectural expressions:

> This tradition of thought goes back to the romantic era with its heavy emphasis on sentiment (rather than intellect), on nature and landscape, on history and on rootlessness. Like Novalis, it contrasted the heroic (and happy) middle ages with the degeneracy wrought by modern times. The golden age, in this view, had existed in the distant past, and it was never quite to be recaptured in the future, for the industrial revolution had uprooted the folk and rural rootedness with all its virtues had given way to urban dislocation —with all its vices.[16]

But this response to the breaking up of rural, preindustrial society may have had its roots in other than nostalgia. As Nisbet suggests, it was also a reply to the anonymity, anomie, anxiety, and general insecurity of the emerging industrial and urban condition which sought to recapture the cohesions and constraints of the small town.[17] Reminiscent of Laquer, he notes that this can lead to totalitarianism in something of the manner of Erich Fromm's "flight from freedom," since people might want to fit into a coherent, meaningful whole at almost any cost. Perhaps it ultimately answers Kingsley Davis's question: Can the anonymity, mobility, impersonality, specialization and sophistication of the city become the attributes of a stable society, or will society fall apart?[18] The organic architects implicitly presumed a falling apart and offered their organic communities as a physical— albeit partial—solution to the problem.

Gilbert Herbert has provided a synoptic view of varying expressions of the organic analogy in the planning of these communities: he says the organic analogy

> . . . derives from the world of living things, from the animal or vegetable kingdoms: it has profound undertones of life—birth, growth, change, ultimately death—in contradistinction to inanimate or inorganic matter. . . . This equation of organic with life is the basic stuff out of which all organic theories are compounded: and the corollary is universally accepted, that the organic plan is the one that fosters the life-process. The bedrock of all organic theories of town planning is a truism, or rather a series of truths held to be self-evident, which may be stated

thus: town planning is the art and science of ordering the environment of communities; organic town planning, by definition, is the process by which a life-enhancing environment is created; the existence, and the further development of civilization is dependent upon a favorable urban (that is, community) environment. Hence, it is argued, organic town planning is a *sine qua non* for the optimum development of civilized man.[19]

Herbert reduces the analogy to five basic expressions: cosmological, natural, systemic, ecological, and cellular: *The Cosmological Analogy* "relates to that fundamental aspect of organic theory which is concerned with the universal problems of inherent order and meaning." [20] The planner is seen here ordering the environment in accord with some larger cosmological or metaphysical scheme of the universe. He seeks knowledge of the structure of the universe in general in order to provide structure for his particular interest—the city. Herbert finds this cosmological bent in both Wright and Gropius.

In *The Nature Analogy,* "the organic is equated with the natural." [21] The organic is in processes, particularly those of living and growing systems as opposed to that which is static and/or artificial (man-made). Herbert considers this argument to be in reaction to early mechanistic thinking and its consequent urban product: the nineteenth-century industrial town. However, he does not relate it to those one instinctively thinks of in this regard: Morris, Geddes, and Mumford.

*The Systemic Analogy* leads the planner to look for "system" in animal and plant life, and results in the planner's organismic vocabulary. It is the analogy of elements interacting abstractly in "systems" and concretely in "organs." Herbert finds it largely restricted to traffic engineering and other limited aspects of planning.

*The Ecological Analogy* is the one with which Mumford and Geddes are associated. It is concerned with the problem of symbiosis in attempting to discern the "nature of the urban community as it is affected by the city." [22]

*The Cellular Analogy* "arises partly from the consideration of the form of society as organic, and partly from the consideration of natural organisms as cellular." [23] Herbert maintains that this analogy is at the root of Perry's neighborhood unit, Stein and Wright's Radburn, and Le Corbusier's superblocks.

Herbert seems to miss the essential point that all of these expressions are essentially concerned with some type of system. All are, in fact, holistic views within which the city is viewed as a system of interdependent parts. Some orientations are more romantic than others in their emphasis upon natural or biologic systems as the appropriate model, but all ultimately depict systems.

While Herbert was primarily interested in the expression of the systems idea in planning, Bruno Zevi has explored its architectural manifestations. In doing this, he also takes an entirely different tack in attempting to define it. In contradistinction to Herbert's orderly, systematic breakdown, Zevi suggests a somewhat amorphous dichotomy. There is again a strong association of the idea with Romanticism, as he sees the organic to be a creative act in contrast to an inorganic intellection.[24] He uses Goethe's distinction between a "fine" versus a "formative" art to support his argument. Although Zevi makes no reference to it, one instinctively thinks of his dichotomy as formulated around a Nietzschean "Dionysian" (frenzied, spontaneous, organic) act contrasted with an "Apollonian" (measured, orderly, inorganic) intellection. To contrast these two perceptions, Zevi borrowed heavily from Walter Curt Behrendt; he developed a concise paradigm of the dichotomy: [25]

| **The Organic** | **The Inorganic** |
|---|---|
| Formative Art | Fine Art |
| Product of Intuitive Sensations | Product of Thought |
| Work of the Intuitive Imagination | Work of the Constructive Imagination |
| | |
| In Close Contact with Nature | Contemptuous of Nature |
| The Search for the Particular | The Search for the Universal |
| Delighting in Multiformity | Aspiring towards Rule, Order |
| Realism | Idealism |
| Naturalism | Stylism |
| Irregular Forms (Medieval) | Regular Forms (Classic) |
| The structure like an organism that grows in accord with the laws of its existence, with its own *Specific Order* in harmony with its environment, like a plant or any other living organism | The structure like a mechanism in which all the elements are disposed in accord with an *Absolute Order* in accord with the immutable law of an a priori system |
| Dynamic forms | Static Forms |
| Forms based on freedom from Geometry | Forms based on Geometry and Stereometry |
| Product of Common Sense (native architecture) of reasonable beauty | The Search for Perfect Proportion for the Golden Section and for Absolute Beauty |
| Anti-composition | Composition |
| Product of contact with Reality | Product of Education |

Since Zevi relied to such a large extent on Behrendt, perhaps we had best have a brief look at him ourselves. His concept of design is relatively simple:

Whether regular or irregular, static or dynamic, all form is a final result of the desire for order. To build is to make a plan. To plan is to follow a definite concept of order.[26]

It is when he delves into the question of forms being regular or irregular, static or dynamic that Behrendt's distinction between the organic and inorganic qualities of design becomes evident. For example, he contrasts the organic and medieval Castle of Nuremberg with the inorganic and Renaissance Palazzo Strozzi in the following terms: the Castle of Nuremberg is irregular in form and does not seem planned at all; it has grown wild. It is "built up by the dynamics of nature that continue to act on its structure." [27] The idea of process with which the social organicists were so fond becomes relevant here in that the building adapts itself over time to the natural conditions of the site on which it is situated, "like a plant that draws its nourishment out of its environment, out of the accidents and conditions of its existence . . . this building seems in the act of adjusting itself to its lifespace." The building provides an adequate expression of the functions it serves, those of shelter and defense, in fittingly irregular forms. The irregularity of the structure is an inevitable result of the process by which each part of the building is adapted over time to the particular functions it serves. The final form is thus like all forms of organic growth: complete, coherent, yet full of individual character with each part of the design absolutely interdependent upon the other parts.[28]

In contrast, the Palazzo Strozzi is complete regularity. Its formal rooms are grouped about a rectangular system of axes which results in a structure of simple cubes. Its facades are symmetrical with a regular division of the walls accomplished by a grouping of windows at equal intervals. It has the same functions as Nuremberg, shelter and defense, but these functions are suppressed in formal expression to a general law of regularity which pervades the whole; a law of geometry and symmetry. The result is static form derived from abstract principles of order. Behrendt further extends this idea by saying that when the ordering involved is that of the city, the planner who thinks "inorganically" demands a leveling of the site in order to impose this concept of regularity.[29]

A varying perspective on use of the organic idea is provided by John Burchard and Albert Bush-Brown, who find in organicism

a belief that the universe resembles the organization of a plant or animal, rather than a machine. This view has natural alliances with [formalism, romanticism, mechanism]. . . . It shares with romanticism a belief in highly individualized living things which are uniquely adapted to their particular environment. It shares with mechanism the belief that physical laws describe the actions of inert bodies and that structure should be visible. It has less that resembles formalism, but its emphasis on the primacy of organic forms is similar. The organicists suggest that evolutionary theory, including adaptation and natural selection, gives hints as to how buildings should be organized. Hence, they insist that buildings be adapted to their environment, both physical and social, to their sites, uses, materials, authors, even to the nations in which they stand. Frank Lloyd Wright's thesis that form must change with changing conditions is anathema to formalists and mechanists or any who believe in the universality of particular geometric forms.[30]

As a final point in our general definition, it should be noted that the organic approach to design is "scaleless" with respect to the size or scope of the object designed; as an approach to design, it applies equally to the macrocosmic city or region and to the microcosmic art object. Some of the associations that immediately come to mind are Richard Wagner's idea of "total art" and Frank Lloyd Wright's idea of "integral" art.

Despite the protean character of architectural use of the idea, two elements dominate: system and nature. Its use is generated by a search for systemic structure at scales varying from the macrocosmic (cosmological analogy) to the microcosmic (cellular analogy). There is a stipulation that the object designed—to which a concept of order is to be given—must have a holistic character and, as a result, there is a search for a concept of system to relate parts within a coherent whole:

A design may be called organic when there is an harmonious organization of the parts within the whole, according to structure, material and purpose.[31]

The architect has observed this systemic quality in nature where parts may have individuating functions resulting in specific "fitted" forms, yet all are bound up in an ordered whole. Whether he has looked for system and found nature or looked at nature and seen system is a moot point given the romantic and obscure nature of the literature.[32]

The best exemplars of this approach are Louis Sullivan and Frank Lloyd Wright, both of whom made of organic architecture a *cause célèbre.* Sullivan's interest in the natural and his searching for principles of order in nature may, in themselves, be only a reflection of nineteenth-century tendencies. He was strongly influenced both by Herbert Spencer with respect to social thought and by Emerson and Whitman in a more poetic sense.[33] This influence can be sensed throughout the *Kindergarten Chats:*

We, in our art, are to follow Nature's processes, Nature's rhythms, because those processes, those rhythms, are vital, organic, coherent, logical above all book logic . . . gifted with spiritual insight, should use those faculties to give to our art a power, a vital, a creative beauty, that shall make with Nature a harmony and not a discord.

[He says this art] must be an organism—that is, possessed of a life of its own; an individual life that functionates in all its parts; and which finds its variations in expression in the variations of its main function, and in the consequent, continuous, systematic variations in form, as the organic complexity of expression unfolds; all proceeding from one single impulse of desire to express our day and our needs: to seek earnestly and faithfully to satisfy those needs. To make our world a pleasant place.[34]

Sullivan supplements his theory of art as an organism possessing a life of its own, in an elaboration of his famous dictum that form follows function, by asserting that form is not just organization but an "organic" organization: form is determined *because* of function.[35] And so, as in its use in social thought, the idea of function is again an essential component of the organic-systems idea.[36]

Wright expanded Sullivan's organicism into something more in line with Zevi's subsequent characterization of the thought.[37] At the broadest level, Wright asserts that an organic architecture is a product of an organic society; architecture being the physical reflection of a society's functions and aspirations; all is system—everything bound together in a series of interdependencies. Thus, if one would relax the constraint of materials in the following statement, there results as succinct a definition of the organic approach as can be found in Wright:

Now there can be no organic architecture where the nature of . . . materials is ignored or misunderstood. How can there be? Perfect correlation, integration, is life. It is the first principle of any growth that the thing grown be no mere aggregation. Integration as entity is first essential. And integration means that no part of anything is of any great value in itself except as it be an integrated part of the harmonious whole.[38]

Mumford repeats some of the foregoing themes. He identifies the organic concept with a wholeness or high degree of integration in a society and with a corresponding wholeness in the life of an individual living within that ideal state.[39] In this organic society all action is directed toward public ends as corporate, organismic aims for the benefit of the community as a whole as opposed to the simple pursuit of self-interest.[40] In a series of successive judgments, the city grows by accretions and the planner must move along a very loose connecting thread from one means-end relationship to another, with overall objectives not too clearly defined at the outset.[41] The formal result of this process is something on the order of a medieval town, about the organic utility and beauty of which, he, Camillo Sitte, and Eliel Saarinen are enthusiastic.[42] This city is limited in size, since "organic phenomena have limits to growth and extension," something he finds lacking in the contemporary, endless, and amorphous city.[43]

There is an interesting similarity between Mumford's thinking and even his style and that of Alfred North Whitehead. For example, Mumford sounds very Whiteheadian in finding that the city is a "process in being" and thus "no organic urban design for any larger urban area accordingly can be complete once and for all," since it requires fulfillment, "a dimension that no single generation can supply; it requires time."[44]

Walter Gropius is usually not thought of as being of the organic school. If nothing else, his International Style forms would seem to preclude such classification. He evidently valued such an identification, however, as he described Gilbert Herbert's study of his work as the best general presentation of his thought, and Herbert describes him as a model organicist.[45] There is a strong tie here, as there is with Mumford, to the organic philosophy of Whitehead. To Gropius, design is organic by its very nature; its purpose is to unify, to synthesize, to find a "whole" which is greater than the sum of its interdependent parts. He designed this whole from cellular parts in "a reorientation of town-planning, based on a progressive loosening of the city's tightly woven tissue of streets by alteration of rural and urban zones and a more organic concentration of the residential and working districts."[46]

Constantinos Doxiadis's theories also fit within the Wright-Mumford tradition but he has gone them one better: he has renamed organicism. To him it is the science of *ekistics;* the study of settlements as living organisms. In *ekistics*

> the city is never a static monument. It is a dynamic, living organism, being born, growing, decaying, dying, perhaps growing again.[47]

Consistent with this, he thinks that static plans are no longer workable, and that they must be replaced by more dynamic "programs." Biomorphism still finds its place; the hexagonal honeycomb is the ideal pattern of the community. This, of course, closely compares with Wright's early statement that the hexagon is the most natural form for human use. There are five basic elements in Doxiadis's conception of the social organism: man, society, functions, nature, shell.[48] In spite of his peculiar vocabulary these elements nonetheless combine in the now familiar holistic or systems view of society and the city.[49]

To balance the presentation of this organic web in the history of the systems idea, it should be noted that current use of the concept by planners largely ignores the early organic antecedents of the idea. Perhaps it is merely anti-historicism, but architects and planners seem to be largely unaware of the literature that abounds on the idea.

In summarizing architectural and planning use of the systems concept, a reasonable premise is that the early organic thinkers in these fields were rational problem-solvers who required some sort of theoretical guide in their giving form to buildings and cities. In that context, the systems idea was attractive. Biology was a dominant science in the nineteenth century; Spencer and Darwin were household words. At the same time, architecture was going through an axiological crisis or transformation of its value structure.[50] Industrialization had introduced the society to new forms and new functions. The architect could no longer rely on the simple adaptation of existing forms within a specific stylistic institutionalization to meet new functions. Industrialization and the consequent galloping urbanization resulted in a denunciation of "mere aestheticism" as a basis of design, and the architect was forced to abandon eclectic models and look around for a new formulation of method or approach through which he might "give order" or design if he were to plan cities. This was an occasion of enormous significance in architectural thought and one which has been very much under-played in histories of the Modern movement. A very good case could be made for the idea that the architect is still searching for this thought-model—as evidenced by the neo-eclecticism prevalent today.

The organic concept becomes relevant in that the surfeit of nine-teenth-century eclecticism corresponded in time with a period of much thought about "nature as a system" in the social and political sciences and in philosophy. As Carl Becker[51] has pointed out, the eighteenth century was characterized by a prevailing metaphysic or cosmology which worked toward the de-deification of God and the substitution of Nature as surrogate. This naturalism was seminal to later expressions in sociology as reflected in Spencer, Comte, et al. Correspondingly, nature seemed to provide the architect with his newly sought structuring principles: nature had order, nature had system, nature was built up of holistic entities constituted of functionally differentiated yet interdependent parts: in sum, nature provided models of form. Thus, the architect, during Romanticism, returned to an ancient aesthetic dictum: follow nature.

This natural organic model allowed for a functionalism dynamically conceived yet premised upon the assumption that form does follow function. At an elementary level, the relationship of function to form can almost be viewed as a simple one-in-one correlation; take, for example, the common table knife. Essentially, its functions are two-fold: to provide both a cutting edge and a means of being grasped comfortably by the hand. The forms these two functions generate—despite the demands of stylistic assertiveness and the capriciousness of fashion—are simple and relatively constant: essentially a single stimulus resulting in a single response. There is no network of stimuli demanding an integration of response. Consequently, the table knife does not generate a conception of function as process; there are no interrelated sets of acts over time which would require an inter-related—integrated—response. An organic expression at this simplistic level would involve only the integration of the two related but separate forms generated by the two related but separate functions within a harmonious whole. As a result, we have the simple linear artifact composed of blade and handle. But when the functions to be housed are processes—a linked series of events occurring over time—the problem becomes much more complex; obviously, any organic view of the city must be one of interrelated processes or systems. This accounts for the seemingly excessive interest by architects in biological systems: they provide an example of how nature houses processes and gives form to the container which envelops them.

Thus the planner, and especially the architect, have often looked rather closely, perhaps too closely, at nature for their organic structuring principles: their models of a systemic integration of parts into a whole. A few have rather simplistically sought a purely formal correspondence between their own and natural forms with little or no thought given to the functional determinants of the forms.[52] They simply established a surface or volumetric simulacrum and felt their task had ended. This superficial use of the organic analogy has resulted in the unfortunate confusion associated with the model and has been the source of much of the derision directed at it. This facile use of the analogy is most evident in Soleri, Doxiadis, and Neutra. Neutra, for example, simply asserts that the city is "in the first place a phenomenon to be understood on a biological basis." [53]

Despite excesses in the use of the organic idea, we cannot be too critical of all the biologic undertones in architectural use of the analogy, as it appears to be a pervasive association. Zevi, for example, found this biological viewpoint in Vasari, Michelangelo, Geoffrey Scott, and Arnold Whittick.[54]

It is, however, the biological conception of system that has generated difficulties in vocabulary which obscure the historical development of the systems idea. The issue originates in Plato's day when there was a tendency to describe the systems idea in terms of natural or organic systems. The early models of system were often expressed in a correspondingly biological vocabulary and imagery since these were the best known types of system and, while there are many characteristics which differentiate modern systems theory from the older organic idea, historically, expression of the idea through use of biological analogies has remained dominant.[55]

This has led many to perceive biological associations in areas far removed from that field. Consequently, some contemporary sociologists would like to shun any association between their "modern" thought and earlier organic models.[56] This is at least in part a reaction to the excesses of Herbert Spencer and of Social Darwinists generally. Discontent with organicism is expressed in dismissal of the analogy as one which reduces a specific discipline to a mere "biologism" [57] or as one which is fundamentally erroneous because it fails to account for the accumulation of culture which is unique to man and which is thus "superorganic." [58] In the extreme, it has even been termed "an absurd and childish" analogy.[59]

To balance the account, however, it should be noted that some "modern" systems thinkers—general systems theorists and engineers, for example—do not share the sociologists' fears in relating their constructs to organic models. In fact, many cite an organic model as the example *ne plus ultra* of a system. For example, Ludwig von Bertalanffy defines a system as "a complex of elements in mutual interaction" and follows it up immediately with the statement that "every organism represents a system." [60] The organic analogy is apparently found appealing in these areas because it is a suggestive heuristic; something that has potential relevance to our field. As cybernetics expert W. Ross Ashby says:

> Cybernetics is likely to reveal a great number of interesting and suggestive parallelisms between machine and brain and society. And it can provide the common language by which discoveries in one branch can readily be made use of in the others. [61]

Norbert Wiener, the father of cybernetics, generalized this point when he said that 'at every stage of technique since Daedalus, the ability of the artificer to produce a working simulacrum of a living organism has always intrigued people." [62] Not surprisingly, systems engineers have a similar view. Ralph Gibson, for example, says, "A system is an integrated assembly of interacting elements, designed to carry out cooperatively a predetermined function." [63] And while there may not be many biologisms in the vocabulary of systems engineering, the approach as he defines it again retains strong overtones of organicism since, "the living organism, the animal, or above all, the human body with its central nervous system, is an example par excellence of a system." The attractiveness of this exemplar of a delicately integrated system is expressed in Gibson's account of the human body as system:

> The human body has been systems engineered by Nature through processes that have taken millions of years and are still going on. Although these processes sometimes seem wasteful, dependent for fundamental improvement on change breakthroughs in the form of new species with greater survival qualities, the results command our utmost admiration. Economy of materials and flexibility of design characterize the structure which houses and implements a brain and nervous system capable of receiving and interpreting intelligence from the external world and controlling the reactions of the body with delicacy and precision. The whole is powered by an alimentary system where energy from chemical sources is supplied upon demand through a fascinating chain of chemical processes. It is an integrated assembly of thousands of elements all cooperating to carry out a certain function, namely, survival in a competitive and changing world. In this system, the elements are essentially simple but so numerous and interdependent that the system as a whole presents a very complex picture to the observer. It is so complex that for centuries patient empirical studies in the healing arts, and later scientific research in medical sciences, have been necessary to support the activities of the practitioners of medicine and surgery who really act as repairmen to the system. [64]

Distinguishing between an "organic" entity and a "systemic" one is a question which occurs with distracting frequency in the history of the systems idea. There must be an appreciation of the fact that the organic and the systems idea are essentially the same; the significant distinction being that organic thought has focused on a particular type of system—a living one; whereas the systems idea is larger in scope —concerned with both living and nonliving systems both conceptual and real. Confusion is generated because of a historical tendency to identify the biological functions of living organisms with alleged characteristics of human society. While occasionally suggestive, this has more often been a gratuitous transferral of ideas: a reductionism. It has also been misleading, particularly in the case of Spencer. Becker and Barnes identify at least three important distinctions between biological and social systems:

> In the first place, in an individual organism the component parts form a concrete whole, and the living units are bound together in close contact, whereas in the social organism the component parts form a discrete whole and the living units are free and more or less dispersed. Again, and even more fundamental, in the individual organism there is such a differentiation of functions that some parts become the seat of feeling and thought and others are practically insensitive; whereas in the social organism no such differentiation exists; there is no social mind or sensorium apart from the individuals that make up the society. As a result of this second difference there is to be observed the third distinction: namely, that, while in the organism the units exist for the good of the whole, in society the whole exists for the good of the individual members.[65]

Because of these distinctions, it has been suggested that the idea of a system in a biological sense be called organismic or bio-organic while in the larger nonbiologic use it might be called organic or simply systemic.

In his use of the systems idea the architect was looking for models of form and, as a result, he focused his view upon natural systems since they were, especially, processes embodied in physical containers: in actual forms. Thus, his problem and his thinking were really of a different type from that of the social theorist. This fact has been largely overlooked by critics who would like to dismiss the architectural stream of thought as either metaphysical, romantic, or just plain nonsense; a characterization that overlooks the powerful heuristic value of the idea in the minds of men like Wright and Gropius. As Modissette pointed out, the architect went through a value crisis and, as a result, sought new models of form. When we relate this to Lovejoy's idea of the social "ambiance" of ideas, the focus upon nature as a model of form seems a reasonable development occasionally carried to excess.

40

Perhaps the most useful illustration of how the idea is used by the architect is expressed by Frank Lloyd Wright. He is not searching for a superficial similarity between the forms he creates and those of nature as do Soleri and Doxiadis; rather he looks to what forms nature developed when confronted with the same problem he has—how nature gives form to specific functions. For example, take the common architectural problem of joining a beam to a column:

> For the architect what a marvel of construction that sahuaro! Or the latticed stalk of the cholla! Nature, driven to economize in materials by hard conditions, develops, in the sahuaro, a system of economy of materials in a reinforcement of vertical rods, a plaiting of tendons that holds the structure bolt upright for six centuries or more. Study the stalk of the cholla for a pattern of latticed steel structure or the structure of the stem of the ocatillo that waves its red flags from the tops of a spray of slender plaited whips 15 feet long.[66]

Other planning theorists have formulated their organic constructs similarly, with specific biologic or natural models. One immediately thinks of the influence of Geddes on Mumford. Others developed this approach independently, although still within the influence of eighteenth-century naturalism and nineteenth-century biology. For example, John Bookwalter thought urban form could be generated by a biologic model. Written in 1910, his definition of the organic relationship has a Sullivan-Wright ring to it: "the principle of organic life is the just correlation of integrant parts having a due and full exercise of special and appropriate functions." When Bookwalter extends this to bio-organic laws of urban growth, it strikingly presages Mumford's thought: [67]

—All organisms, by virtue of the conditions under which they exist, tend toward new forms;
—Violent action is harmful to a sound and stable body;
—A growing organism is composed of cells, all of which are located in their proper relative positions;
—The organic body grows not by the enlargement of the cell but by the multiplication of the cells;
—The organic body must possess flexibility in order to promote its growth;
—The organic body has within its capabilities the ability to establish the relationships of cells.

The organic idea found didactic expression in that D'Arcy Thompson's *On Growth and Form* was occasionally required in design courses. Paul Jacques Grillo, an architect, has discovered formal evolutions to be roughly similar in natural and man-made systems having the same functional infrastructure.[68] The two systems he relates quite cogently are aircraft ·and fish, both of whose forms have developed in response to the specific functional demands of the laws of fluid dynamics. This recalls early architectural enthusiasm with aircraft and steamship forms by Le Corbusier and others.

There is a tie here with systems engineering in that a study of shark and dolphin forms, with respect to fluid flow characteristics, preceded hull-design studies of the prototype submarine *Albacore*. Another instance in the same area is the esoteric systems specialization of bionics—a shortened form for biology-electronics—where the structure of a frog's eye provided a conceptual model of an advanced radar system. Interesting also is the definition of bionics: "the science of systems and devices which function in a manner characteristic of or resembling living systems," or what Wright said architecture was.

Perhaps the most significant expression of this thinking today is that of the so-called "Metabolist Group" in Japan who have adapted certain biologic concepts to urban planning. Their conceptual scheme is based on "change, growth, interaction, and simultaneity of functions." They perceive a conceptual model in the continuous mutation processes of nature and as a result have changed their orientation from one of master planning to systems planning; their conception of system being influenced by natural metabolic and growth patterns exhibited by natural systems. In sum, they see their task as one of relating form to a continual modification of the whole and its constituent parts; [69] the parts being interdependent or mutually supporting systems of the city.[70]

This usage of the organic concept has retained the essentials of our earlier consensus view. Organicists look for system in a holistic entity and they observe that a model exists in nature; specifically, in biological systems. Theories based on the idea bear different labels: "integralism" or "total architecture" (Gropius), "ekistics" (Doxiadis), or the usual "organicism" of Wright and Mumford; but essentially these are similar approaches to the problem of designing a systematic whole, be it a building or a city. These various theories can best be viewed within a polar frame of reference. We may postulate an organic (systems) pole at one end and an organismic (biologic) pole at the other, with something of a theoretic continuum between them. At the organic pole we can visualize Gropius and Giedion with their abstract "integralism," in search of a harmonious synthesization of urban processes within an integral whole; biology is absent in their abstractions. At the other, the organismic pole, we have Soleri and Neutra with their "biologism"; at varying points along the connecting continuum, we find Wright, Sullivan, Mumford, and others.

In this context, organicism has been an approach, perhaps a "movement," in design. It is not a style as it must allow for the formal diversity of, minimally, Wright and Gropius. As an approach it is characterized by an emphasis on the holistic nature of a problem; of viewing a work of architecture or a city as a whole composed of functionally differentiated and interdependent parts. Function is conceived of in terms of processes and a model is sought in nature due to nature's integration of diverse processes in functionally efficient forms. It seeks a comprehension of this natural translation of function to form. It is tied to nature by other means also as it is saturated with the biosocial views of LePlay, Geddes, Spencer, Cooley, et al. It is also highly charged with Romanticism and, as additional baggage, carries with it an association with some controversial ideologies.[71]

This has led some to dismiss organicism as metaphysical and romantic; a Mumfordian "blurry blueprint" of little value,[72] premised upon "utopian logic, physical environmentalism and [a] custodial view of the community's future."[73] No doubt there is some truth in this. Perhaps an even more damaging indictment could be made to the effect that there is an enormous cultural lag in architectural thought: Spencer used the analogy with similar simplistic biological overtones over a century ago and, as a result, it fell into disfavor; architectural use began with Sullivan and the criticism has yet to be fully articulated.

But, while in some ways valid, this facile dismissal is itself at times superficial. The architect and physical planner are concerned with giving form; with providing a container, a shell, for complex processes. They have recognized that these processes operate interdependently —as a system. Others have not. As Paul Goodman pointed out:

> The policy of a modern city is worked up by its highway engineer, its houser, its sociologists, its school superintendent, its tax expert, its political administrator, each in his expertness. When the whole is then put together, it comes to delinquency, traffic congestion, crashing civic ugliness, and these too are worked on as special problems, with new levels of administration, ad hoc programs for dropouts, face-lifting, one-way streets, and—needless to say—new millions of dollars for new experts. Nobody thinks about the *community*. To counteract this kind of scholasticism, our mid-twentieth century academies must be places where learned specialists can temporarily suspend their beautiful methodic skills that necessarily define too accurately and exclude too much.[74]

It is the so-called organicist in planning that has focused on just this idea of community as whole or entity. While this may have resulted in occasional nonsense, it has also brought about an interest in the interrelatedness and mutual dependencies extant in the city—the city as a system. Hopefully, therefore, with some assists from the more "rational"—or at least less romantic—disciplines, the organic-systems concept will continue to suggest insights into the complex process of planning and design.

# II
# Use of
# the Systems
# Idea

We can now move on to a look at actual use of the systems idea in architecture and urban planning. This illustration of use will be selective. Our criteria of selection are either a) a particular use or expression that is quite significant in practice and consequently must be accounted for, or b) a use that is especially descriptive of the idea's basic content of holism and rationality. Our aim is ultimately an assessment of the utility of the systems idea of architecture and planning. In that context, little technical detail can be developed but hopefully our account will be sufficient to provide the reader with a sense of the idea's potential. In sum, while an encyclopedic catalog of real or potential usage is not possible, a brief account should lead to a fuller understanding of the idea's content and usefulness.

## 6. The Systems Idea as a Guide to Assessment and Choice

**Introduction.** Ideally, the planner or designer, in assessing potential or alternative projects, would like to know with some precision what probable stream of consequences might flow from a project over time. Moreover, he would like to identify the individuals and groups that would be the recipients of these consequences and to classify the consequences as either positive or negative ("benefits" or "costs") as perceived by the probable recipients as well as by himself. A possible classification of consequences or "effects" of the project could be: social, economic, political, physical, psychological, and miscellaneous, depending upon the nature and scope of the project. For those consequences that are economic in nature, the planner wants to know the precise financial outcomes. For all consequences, he would like at least some measure of the probable perception and reaction of recipient groups in assessing a project, i.e., to what degree the final outcomes or consequences are significant as they are perceived within their varying value systems.

To arrive at this assessment (i.e., to evaluate a project), what is needed is a methodology or set of techniques that will identify the consequence pattern generated by a project, its reception by a defined public, and some weighting device or general "welfare function" which allows the decision-maker to choose between that which is "good" or "bad" with respect to defined criteria, be they social, aesthetic, economic, or whatever. Realistically, at a city-planning scale, the choices are between alternative candidate projects for funding, and the decision is one of determining the more meritorious projects with respect to some rather nebulously defined public interest or social good.[1]

Problems of assessment and choice have been of sufficient interest in the planning of public projects to have generated a number of mandates either from the Congress with respect to cost-benefit analysis or from Presidents with respect to a planning, programming, and budgeting system.[2] The result is a whole range of techniques that have been developed in an attempt to assess the consequences of an act (read also: program, plan, project).[3] In most instances, these techniques are rather complex and demand a good deal of technical analysis. Our concern will be only with those techniques which have found specific applications in architecture and urban planning and, to limit the scope even more, to brief, nontechnical illustrations of the techniques. Our presentation will range from simple to more complex methods—from those that are rather precise but attempt only a limited assessment of consequences to those that try to encompass more of the real impact of a project at the cost of less precision. Specifically, from engineering economy studies and cost-benefit, cost-effectiveness analysis to the planning balance sheet, and the goals-achievement matrix.[4]

**Engineering Economy Studies.** Cost-benefit analysis, as it has been traditionally used in architecture and planning, is really a summary term for several techniques that assess the relative economic utility of alternative projects or schemes. In architecture, the term "cost-benefit analysis" is something of a misnomer, the currently fashionable term being "life-cycle costing," but generically the techniques have been traditionally known as "engineering economy" studies.[5] These methods are limited to an assessment of economic consequences only and collectively they involve the discounting of various cash flows—the expenditures and receipts or dollar "costs and benefits" of a project over its probable life span. All provide a comparable or equivalent basis for financial evaluation. The four principal formats for such analysis are:

**1.** Present worth analysis.
**2.** Equivalent uniform annual cash flow.

**3.** The benefit-cost ratio.

**4.** The prospective or internal rate of return.[6]

While the four methods vary with respect to technique and more importantly in the end product of the analysis—a requirement established by varying decision contexts—all are essentially similar in principle. Only the first three will be examined here as this should be sufficient to elucidate the principles involved.[7]

The significant characteristic of these methods is their accounting for the function of interest in the flow of money over time. In the planning of private projects, the test of financial feasibility is obvious: to an investor, a dollar received now is more valuable than a dollar five years hence because the dollar now in hand can be invested at interest to generate a larger sum five years hence. Thus, the investor is interested in the return on his investment, a return which must be determined with a certain precision. More importantly, when confronted with two or more alternative projects, the investor wants to know which will have the greater probable yield. In this context, the techniques are also known as financial or capital expenditure analysis. They are adopted for public projects (as engineering economy or cost-benefit studies) on the presumption that only the more attractive projects should receive funding from the public purse. Generically then, these techniques portray the expenditure and receipt of monies over time at an appropriate interest or discount rate so that alternative projects can be assessed on a comparable basis.

A useful instrument in making these assessments is the so-called cash-flow diagram (Fig. 11). This depicts the flow of monies from a project over time—both receipts and expenditures. Three variables are accounted for: vertically, the benefits (or receipts) and the costs (or disbursements); horizontally, the distribution of both over time. The conventions are of course arbitrary, but in our notation, those funds illustrated above the horizontal axis are the positive cash-flow (receipts) and those below the negative cash-flow (expenditure). It is also important to note that cash-flow diagrams illustrate a specific point of view as to who is receiving and expending the money. Figure 12 illustrates borrower and lender's cash-flow diagrams for the same loan.

Any meaningful economic analysis must account for the fact that a dollar today is worth something more than a dollar at some future time and, conversely, money to be received later in a project's life is worth something less at the present time. In the present-worth method the entire stream of receipts and disbursements which is anticipated over the life of a project is discounted back to the present or to a defined starting point for the project. This allows all costs and benefits to be compared as discounted to a common moment in time at a specific interest rate. In Figure 13, only a single cost is illustrated at the outset of a project and after the project's life (of *n* years) a single benefit is received. Obviously this example is simplified in the extreme, but it does illustrate the principle that a future benefit must be discounted, via the arithmetic shown, back to the "present" in order to provide comparable costs and benefits. At the "present" time, discounted benefits must at least equal the discounted costs of the project if the project is to be financially feasible at the specified interest rate. It should also be noted that in the real world one does not normally bother with the arithmetic as shown; rather, one refers to a set of compound interest tables, at the relevant interest rate, simplifying the calculation significantly.[8]

An issue that arises in all economic studies of this type is the determination of an appropriate interest or discount rate.[9] Some argue for using the rate one must pay for capital in the market at the time the project will begin. Others suggest an anticipation of interest rates during the life of the project and then an averaging of them out. Some see a significant distinction between interest rates assigned to public versus private projects. Finally, some view the entire issue as a question of value judgments. This determination is complex and, for our purposes, it is sufficient to indicate that an interest rate must be chosen and that the choice of a specific rate can often have a significant effect on the ultimate financial outcome.

Assuming that such an interest rate has been defined, the present net worth is the difference between the present net benefits and the present net costs. If this present net worth figure is positive the project is earning more than is expended on it and is economically viable. If the planner or architect is confronted with alternative projects which are somehow mutually exclusive, an analysis of the alternatives by the present net worth method, discounted at the same rate, results in three comparable present net worth amounts, the greatest of which identifies that project with the most desirable economic outcome.

The present worth method is illustrated more graphically in Figure 14, which represents an example that is more typical of the real-world building process. In the example an intial cost is incurred and a series of uniform annual benefits is received. There remains a simplification in the assumption that all design, legal, and construction costs occur instantaneously at a single moment (the "present") and that the benefits the building will yield (as rents) are absolutely uniform in their distribution over the project's life. These constraints are removed in the final and more general example (Fig. 15) which depicts an irregular series of cash flows over the life of a project.

In the planning and design of public as opposed to private investment projects, it is often more useful to have the flow of monies on a project distributed in a series of equivalent uniform net annual flows. This allows for a better integration of project financing with public budgeting processes which are normally on a fiscal year basis. In such cases, the present net worth of a project, either positive or negative, can be determined as in previous methods and then redistributed in an equivalent uniform net annual series (Fig. 16). Again, in practice the arithmetic is not usually done as shown here. Rather, a financial table is consulted for an appropriate "capital recovery factor" at a specified interest rate and period of years which is then multiplied by the present net worth to determine the equivalent net annual stream of income or expenditure that the project will generate. Ultimately, the values derived through those techniques can be converted to a benefit/cost ratio of present net values or, alternatively, of equivalent uniform annual cost and benefits. This is illustrated in Figure 17, which shows an irregular series of costs and benefits over the life of a project converted via the previous methods into the present worth of costs and benefits from which a benefit/cost ratio is determined. The utility of this technique is that a unique numerical measure of rating projects via the ratio of benefits to costs can be determined. Again, the presumption is that both benefits and costs have been discounted over the project's life at a comparable rate of return and, at that rate, the ratio provides a defined measure of economic utility: those projects in which the ratio exceeds one are economically viable, and the greater the positive value of this ratio the more desirable is the project.

**Cost-Benefit Analysis.** Our illustrations so far, while having a certain utility in traditional architectural and, particularly, engineering practice, are much too simple for the planning of public projects in which the consequences involve more than simple fiscal feasibility. As a result, cost-benefit analysis as practiced at a larger scale is enormously more complex.

It is difficult to determine when cost-benefit analysis was first used. While economists find significance in several early works in their field one of the first books on the actual practice was Wellington's *Economic Theory of Railway Location,* published in 1887.[10] Use of the idea in public planning and design was, until recently, largely restricted to water-resource projects under the purview of the U.S. Army Corps of Engineers. For the planning of these projects, Congress issued various prescriptions such as the Rivers and Harbors Act of 1902 (with updates in 1927 and 1928) and, most significantly, the Flood Control Act of 1936. In this act, a Congressional mandate was established to allow federal participation in projects only "if the benefits to whomsoever they may accrue are in excess of the estimated costs." [11] Cost-benefit analysis in planning was thus institutionalized and there the complexity in use of the technique also began: the planner had to account for all consequences of a project and by virtue of, "to whomsoever they accrue," all recipients of consequences: river users, farmers, urban dwellers, and the like.

As the method developed so did the bureaucracies concerned with it. In 1943 the Corps of Engineers, Department of Agriculture, Bureau of Reclamation, and Federal Power Commission combined to form a Federal Interagency River Basin Commission which established criteria for governmental cost-benefit analysis. Again, the criteria established were largely economic (consequences were measured in dollars and the greater the ratio benefits to cost the more preferred was the project), but ultimately there was a gradual expansion of definitions of "cost and benefit" to today's more ambitious attempts to assess noneconomic consequences such as those of a social, political, aesthetic, and psychological nature.

In general, the more ambitious these efforts (i.e., the greater the scope and content of the consequences assessed), the less successful they have been. This is inevitable given the problems of a more holistic analysis. As economist Samuel B. Chase points out:

> No economist, presumably, would argue against the spirit of benefit-cost analysis. Once implementation is considered, however, problems arise at every level, from pure theory to the generation of "real numbers." At the purely theoretical level, ideally, this approach requires that there be, in fact, some such thing as "social optimality," that there be such a thing as a social welfare function, and that this social welfare function have certain technical properties. Economists have yet to reach a consensus on these matters.

> Beyond this, it is necessary to develop procedures for quantification. Granted that there are social costs and social benefits, how can they be measured? Given an accepted conceptual approach to this measurement, what are the prospects for developing accurate estimates? How far does it pay to go in making improvements in estimating techniques? How much faith can be placed in the results of such studies? [12]

In resolving these problems, analysts are often preoccupied with such questions as: 1) Why, in fact, should future social benefits and costs be discounted at all? Should one pay an interest rate on social or political consequences anticipated in the future as one does on purely monetary returns? 2) If cost and benefits are to be discounted what is the appropriate discount rate? 3) Is there a generic social discount rate or should it vary from project to project? Should projects with greater risks or those of an experimental nature have higher discount rates than those of a routine nature? 4) How can one identify the distribution of benefits and costs which will flow from a project as ultimately received by different groups within a given community? This is known theoretically as a redistributory impact analysis and it generates such questions as to why Group A should pay for benefits to be received by Group B. 5) Should certain compensation mechanisms be built into a project if Group A is in fact to pay for benefits received by Group B? 6) Perhaps most significantly: how does one determine a quantifiable measure of those benefits and costs that are not reflected in market prices? 7) Also: how does one really assess the political and social consequences of a project, and then how can these assessments be incorporated within an overall project evaluation process? [13]

Although superficial, perhaps, this accounting of the problems involved does indicate the types of questions that enter the minds of economists when they are thinking about cost-benefit analysis. The questions are not as abstract and theoretical as they may appear. For example, a very real problem is how does one account for those undesirable characteristics of a project which are not intended; so-called "adverse effects." Should they be considered as costs pure and simple or as negative benefits; so-called "disbenefits"? This is important, as inferred in the preceding example (Fig. 17); whether one adds them on to the costs or simply subtracts them from the benefits will affect a benefit/cost ratio:

> The phenomenon . . . is one example of a class phenomena which are sometimes called *"disbenefits" from projects, i.e., an undesirable side-effect.* There are two distinct ways of classifying such items, the first being to treat them as negative benefits and place them alongside the positive benefits and the other being to treat them as costs and to place them alongside other costs. The first method of classification is based on the view that it is better to keep all "inputs" (good or bad) separate from all "outputs" (good or bad), while the second method is based on the view that it is better to keep all the good things ("advantages," "arguments for") together (whether inputs or outputs) and all the bad things ("disadvantages," "arguments against") together (whether inputs or outputs). To the extent that this is simply an expositional matter, the choice must obviously be determined by the need for clarity and comprehensibility.[14]

One way to get around some of these difficulties has been a modification of the cost-benefit technique in the form of cost-effectiveness analysis. Michael Teitz has explored this in the context of urban services and says:

> In *cost-benefit analysis* both sides are quantified in dollars and every effort is made to include not just the direct immediate costs and benefits to the agency and its clients, but also indirect or spillover costs and benefits accruing to others not directly related to the action in question.

> Systems analysis did not develop in this context, but rather in the realm of military decisions where the monetary quantification of benefits is largely impossible. Even so, it remains possible to quantify the objectives toward which actions are directed and to devise measures of the cost and effectiveness of alternative actions or sys-

tems with respect to those objectives. If such measures are available, it should be possible to form a good idea of the relative costs of achieving a marginal gain in one direction as opposed to another. *In contrast to cost-benefit, cost-effectiveness analysis allows the use of multiple, not necessarily commensurable, measures on the benefit side.* It does not solve the problem of incorporating all of them into a single benefit measure in dollar terms, but rather allows the decisionmaker to see clearly the cost of achieving specified gains in each of a variety of measures related to any objective. For activities such as social services, in which quantification is difficult and dollar evaluation of direct and indirect benefits subject to arbitrary assumptions, this procedure seems to offer a useful if rocky path toward improvement of information for decisions.[15]

**Planning Balance Sheet.** One of the more successful attempts to incorporate cost effectiveness analysis in urban planning has been Nathaniel Lichfield's *Balance Sheet of Development.* In describing his method, Lichfield says:

The programme for a plan is regarded as a series of development projects, interrelated in time and space, which must be implemented if the plan is to be realized. The programme is visualized as completed and for each project or group of projects are enumerated the parties who will be concerned with producing and operating them (both private and public) and, also the parties who will consume the services provided, whether they buy them in the market or collectively through rates and taxes. For each party is then forecast the costs and benefits that will accrue. Each such item is measured in money or physical terms as far as possible or otherwise noted as an intangible. If alternative plans are being compared, the analysis aims at the difference in costs and benefits that will arise under each alternative. It is possible by this approach to produce systematically, in descriptive and tabular form, a complete set of social accounts which show all the significant costs and benefits which would result from the implementation of the plan, or from alternative plans and their incidence. These accounts can then be "reduced" by eliminating double counting, transfer payments, common items, etc. This is the "planning balance sheet" which provides a summary of the planning advantages and disadvantages to the public at large.[16]

In presenting his planning balance sheet Lichfield provides a tabular array of perceived consequences which accrue to alternative schemes or plans or projects. Moreover, he has developed a scenario in which all actors involved are placed in two categories: producers and consumers (Fig. 18). Consequences are then depicted in three categories with varying symbolism: dollar signs for those which can be expressed in monetary terms; an M if the consequence can find quantitative but nonmonetary expression; and finally an "I" representing intangible consequences. Both M's and I's are subscripted to allow for individual specifications. Finally a dash indicates there to be no consequence whatsoever.

The attempt to account for a project's impact upon the many recipients of consequences in any urban development scheme is as ambitious as it is necessary. And, while one can question the somewhat arbitrary division of consequence recipients into producers and consumers, this technique does have its adherents. Among them are Prest and Turvey, the authors of a classic introduction to cost-benefit analysis. They say:

> Lichfield is primarily concerned to show how all the costs and benefits to all affected parties can be systematically recorded in a set of accounts. The arguments in favour of proceeding this way instead of by simply listing and evaluating the net amount of each type of cost or benefit are threefold. First, starting on an all-inclusive gross basis and then cancelling out to obtain net social benefits and costs insures against omissions. Second, the financial consequences of any project are sometimes important, and it may need to be redesigned in the light of the distribution of benefits and costs so as to compensate parties who would otherwise stand to lose from it, securing their support for it. The notion of compensation is familiar enough in welfare economics; here is a case where it can be important in practice. Third, whether or not financial transfers are used to affect the final distribution of net gains, that distribution will often be relevant to choice. In practice, that is to say, for good or bad reasons, the attitude of a county council or similar body towards a scheme will depend upon who gains as well as upon how much they gain. As the client of the cost-benefit analyst, it will therefore want to know more than just the total net figures.[17]

**The Goals-Achievement Matrix.** Morris Hill has been critical of Lichfield's balance sheet; he questions its validity with respect to its proposed purpose and its usefulness for the evaluation of alternative courses of action. Hill also questions whether or not the costs and benefits included in the balance sheet are relevant, but since this is a criticism that, in some measure, could possibly be leveled at all types of cost-benefit analyses regardless of their authorship it seems pointless to attribute this weakness to Lichfield's technique *per se.*[18]

Hill's alternative to the planning balance sheet is similar in structure. In his "goals achievement matrix," overall goals are expressed (as operationally as possible) in the form of specific objectives; the ideal of achievement being a quantitatively measured increment toward an objective. Again there is a classification of consequences into 1) those that are dollar valued, 2) those that are not dollar valued but quantitatively expressed, and finally 3) intangibles. There is a weighting of objectives to provide the decision-maker with an assessment mechanism which will aggregate costs and benefits. These are compared and reported in separate accounts for each sector of a community and then, as weighted by the significance of the particular objective that they are perceived as instrumental in achieving, they are incorporated in a definition of "the public interest" in assessing alternative plans or actions (see Fig. 19).

Given the range of the previous techniques for cost-benefit analysis in architecture and planning, it is easy to see how one might have a tendency to get involved in the details of the procedure and lose sight of some of the larger issues. While we can agree with Samuel Chase's earlier assertion (p. 49) that no economist would argue against the spirit of cost-benefit analysis, one must also observe that it is precisely the spirit of things that often seems to escape economists.[19] One gets the uneasy feeling in reviewing the literature that many advocates think it is simply a matter of time before we will be able to pin a numerical value on the many nebulous and interdependent issues of significance in assessing a design or plan. But is this really possible? What is one, in fact, attempting to do? Ideally, a responsible and holistic evaluation of alternative projects is nothing short of trying to get desire, frustration, need, symbol, aesthetic, history, morals, economics, politics, and a general sense of the fitness of things neatly arrayed in a reasoned and intelligible analysis. In practice, one often finds that these elements—if they are even considered—are forced into a Procrustean bed in the form of a graph or equation.

Viewed in a somewhat humbler context, the cost-benefit techniques are quite useful—indeed, essential. Decisions must be made and an assessment process is unavoidable. On the urban scale, assessment techniques were traditionally well-embedded (and concealed?) within the political process, reflecting the values (machinations?) of that process. In architecture, it was the great architect or planner telling us the way we ought to have it. While occasionally facetious, Wright's comments about Boston ("you ought to move out and tear it down") or Le Corbusier's prescriptions about the beauties of the right angle and how the city should be shaped by it, were largely dicta from the perspective of architect as ultimate social arbiter. There was really little in the way of developed reasoning in their work which would indicate why one should prefer their prescription over any other. Le Corbusier's right angle at the expense of what he characterized as the pack-donkey trail influenced many architects and planners and undoubtedly led to a good deal of the sterility of the modern rectilinear city plan, to say nothing of the offense it gave to enthusiasts of medieval Italian hill towns who found in them a formal order—an aesthetic, a humane environment—far more subtle and sensitive than any invested in the modern city "engineered" at right angles. Yet, who is to say what the aesthetic "objective function" of the city is to be; such a goal presumably defines the context within which specific urban forms acquire their instrumental character and their cost-benefit attributes. Recognizing the improbable nature of objective definitions of aesthetic goals some might suggest surrogate measures. Here, one can imagine the conjuring up of a Kahn, a Venturi, a Wright index of aesthetic merit to be neatly subsumed within an "urban aesthetic objective function." Would this be more useful than Le Corbusier's "right angle" measure of aesthetic worth?

The difficulties recognized and the absurdities put aside, cost-benefit analysis must be appreciated as providing a limited but reasonable (rational?) context for the assessment of alternative designs and plans. As Prest and Turvey have noted, analysis forces those responsible into a measure of quantification rather than relying entirely on guesswork and also may provide an indication of the potential consumer's willingness to pay. Moreover, a

practical point stressed by some practitioners of cost-benefit analysis is that it has the very valuable by-product of causing questions to be asked . . . which would otherwise not have been raised.

Even if cost-benefit analysis cannot give the right answers, it can sometimes play the purely negative role of screening projects and rejecting those answers which are obviously less promising. This role is akin to that of a University or College entrance examination: that whether or not one can grade the prospective candidates reasonably well on the basis of previous examination performance, personal interview, etc., the special entrance examination at least gives one a cast-iron justification for rejecting the weaker brethren. In much the same way, insistence on cost-benefit analysis can help in the rejection of inferior projects, which are nevertheless promoted for empire-building or pork-barrel reasons.[20]

In this context, the methodology is the reasonable approach of reasonable men attempting, however imperfectly, to find what might be in the public interest (Fig. 20). One can, in fact, foresee the probable evolution of a synthesis of many existing assessment and evaluation techniques. Certainly there will be a role for the traditional engineering economy (or financial analysis) model. A good deal of work is still going on in sorting out the bugs in more elaborate cost-benefit and cost-effectiveness constructs.[21] When this work is sensitized to the operational and political realities implicit in both Lichfield's (p. 51) and Hill's (p. 52) approach, it could be combined with such legally institutionalized assessment techniques as environmental-impact analysis. Thus, notwithstanding the many conceptual difficulties that exist, one can foresee the ultimate production of useful instruments for assessing projects. Whether one called it cost-benefit or cost-effectiveness analysis, the planning balance sheet, or a goals-achievement matrix, is unimportant. The fact is that some instrument of consequence assessment must be available to planners and decision-makers if they are to be at all rational.

The potential for insensitivity reflected in the present "rational" techniques is of course a problem; there must be a concern with something beyond economic impact. In spite of Ruskin's misgivings about living in a city designed by economists, it has been the economists and their engineering economy counterparts who have, in fact, produced the most useful assessment mechanisms to date. These devices have shortcomings but at least they provide partial analyses.

Again, the systems ideal of both holism and rationality is difficult to achieve; as we have seen, the more holistic one becomes the less rational the analysis. Particularly so in the sense of being able to identify with any precision a specific stream of probable consequences of a project and then to order and assess these consequences relative to precision objectives which are themselves instrumental in realizing "the public interest." Nonetheless, the effort continues, the literature in economics on this type of study is burgeoning, and the prospects are hopeful. To it, there must be added the contributions by architects and others sensitive to values outside the normal province of economics, if the assessment techniques developed are to be more than accounts of easily quantified and dollar-valued phenomena.

## 7. The Systems Idea at the City Scale: Modeling Large Systems

**Overview.** A most ambitious attempt at holism and rationality is the modeling of large-scale systems. It is holistic because it attempts to encompass enormously diverse phenomena, and rational because it tries to deal in a calculating manner with almost incomprehensible complexity.

A model, we recall from Part I, is a representation of a system. Such representation is essential because:

No substantial part of the universe is so simple that it can be grasped and controlled without abstraction. Abstraction consists in replacing the part of the universe under consideration by a model of similar but simpler structure. Models, formal or intellectual on the one hand, or material on the other, are thus a central necessity of scientific procedure.[1]

In modeling, the analyst attempts to abstract from the real-world situation or problem those elements—and only those elements—which are perceived as significant and useful in his problem-solving. There is usually an attempt to represent this abstraction as a mathematically expressed set of variables and relationships between the variables. If this quantitative expression of relationships can be established, the analyst's kit of tools becomes most useful; these include such skills and techniques as: Statistics and Probability Theory, Replacement Models, Queueing Theory, Sensitivity Analysis, Linear and Dynamic Programming, Assignment and Transportation Models, Simulation Models, etc.[2] Such tools are essential if an objective analysis of relationships is to be made. The more complex techniques are used in dynamic models: they attempt to define relationships between the variables over time. Models can be descriptive, predictive, or, most ambitiously, normative, i.e., instruments which define what *ought* to be done as opposed to what is or will obtain given no interventions in the system.[3] Some models attempt a combination of these functions.

Given the brevity of this book and the complex nature of most models, they can only be suggested here in a diagrammatic format. Among those interesting to the architect and physical planner are those of cities (Fig. 21), transportation systems (Fig. 22), and housing (Fig. 23). Others have been developed, of course, within many social areas (health systems planning, criminal justice planning, etc.).

The modeling of large systems has been most successful in those areas we traditionally think of as large-scale engineering systems: transportation, water supply, and sewage systems.[4] In these areas, it is usually possible to identify the various elements that affect the system and, moreover, to find a quantitative expression for their relationships. While such systems are simpler and therefore easier to deal with than most encountered in architecture and urban planning, the modeling process itself can be quite complex (see Fig. 24). Some of these models of specific subsystems of the urban system can be linked together in the form of an aggregate model which will describe a portion of overall system behavior (see Fig. 25).

Unlike architects, those of an operations research or mathematical bent—while occasionally indulgent in descriptive or predictive modeling—have normally been reluctant to prescribe for society. One of the few efforts by systems analysts in this area is a model of "Compact City" by Dantzig and Saaty.[5] In examining the urban condition they find that the two most significant variables are space and time. After assessing various historical solutions to urban form, including Howard's Garden City, The Greenbelt Cities, Frank Lloyd Wright's Broadacre City, and Le Corbusier's plans, they suggest a plan for a compact city something on the order of the Garden City refitted within a megastructure intricately designed to respect the critical dimensions of space and time (Fig. 26). Compact City is interesting also in the analysis and design of specific subsystems, for example, health or water recycling (Fig. 27), but it is rather naive architecturally. The essential concept has been presented with greater sensitivity by Kenzo Tange, Yona Friedman, Walter Jonas, Fumihiko Maki, and others. In sum, examples to date suggest that those who excel in the analysis of prob-

56

lems do less well in the architectural synthesis of a solution and vice versa.

**Forrester's Urban Dynamic.** Undoubtedly the best-known model of an urban system is Professor Jay Forrester's *Urban Dynamics*.[6] This model is interesting for our purposes because it attempts to view the city as a whole as well as to account specifically for activity within a physical system: housing. Moreover, the model provides a number of insights and cautions in the use of quantitative models as a means of exploring the urban condition. Forrester feels that existing models are essentially intuitive and much too static: any realistic model must be dynamic—that is, it must account for the cumulative effects of interaction between significant variables over time. The elements of reality that he has abstracted and called "a city" can be aggregated under several categories: geography, variables, relationships, and premises. His perception of the city—highly simplified—can then be summarized as follows:

1. Geography
   a. A fixed land area on a featureless plain.
   b. Crossed by available transportation in 20 minutes.
   c. In its growth phase, occupying an area of 156 square miles (100,000 acres) with a population of 5.7 million people.
   d. A geographic unit that is politically autonomous, which is implied because of the city's assumed fiscal autonomy.
2. Significant variables
   a. Industry, housing, and people; each of which is divided into three classes:
      1) people: manager professionals, skilled, underemployed.
      2) housing: premium, worker, and underemployed.
      3) industry: new, mature, and declining.
3. Relationships between the significant variables are described in a series of rate equations which depict the way components change over time. How, for example, underemployed people (unemployed and the marginally employed) will enter the city at changing rates over time as a function of other variables.[7] A most significant set of relationships, however, is oversimplified in the abstraction: those between the city and its surrounding environment. Forrester sees the significant variables all interacting within the city system itself; relationships—and consequently effects upon the city—with the suburban environment are trivialized. No one, for example, commutes across the boundary.
4. Premises; among the more prominent:
   a. City will be managed on a profit-making or at least on a non-money-losing basis.
   b. Poor people will come into the city and demand housing and services that will ultimately cost the city something in excess of the taxes the poor will pay.

**c.** Middle- and upper-income persons will pay more in taxes than they receive in city services.

**d.** Industries will pay taxes but will not receive services from the city.

**e.** The city will begin on a featureless plain and grow, within 250 years, to a period of equilibrium (i.e., stagnate condition).

**f.** At the equilibrium point taxes will be excessive, a housing shortage will exist for middle- and upper-income groups along with a housing surplus for the lower-income group; there will also be high underemployment.

**g.** It is better to move people with problems (i.e., the poor) out of the city than to solve their problems within it. Where they go is not specified nor is the ultimate solution of their problems; but, the fact that they do leave is perceived as essential to correcting the city's problems.

**h.** Each class of persons will live only in the correspondingly appropriate class of housing.

This brief description does not do justice to the full range of Forrester's vision of the city; in the context presented here it is merely an abstraction of an abstraction. Hopefully, it will help to illuminate the difficulties and potentials in large-scale modeling.

Forrester's model has generated a great deal of criticism, as can be surmised from some of the assumptions listed above.[8] There are elements of unreality and even contradiction between several of the geographic, economic, and political premises. It also supports a Banfieldian vision of benign neglect of urban problems in its suggestion that some current solutions to urban problems are in fact partial sources of the problems.

While a critique can be developed about these points at some length, it is important to note the merits of the model. Forrester has described an intricately structured reality in mathematically precise terms; a series of simultaneous, deterministic, difference equations accounts for the probable behavior of a large, complex, time-variant system. This is no mean accomplishment; while it is yet to be tested against reality, it forcefully illustrates an informed analyst's "counterintuitive" depiction of reality which challenges many of the assumptions of city planners and may ultimately lead to something operationally useful in urban planning.[9]

## 8. The Systems Idea at the Architectural Scale: Systems Building and Building Systems

**Introduction.** Systems building is the architectural profession's response to industrialization. At a time when most social thinkers are concerned with a transition to the postindustrial age, one might wonder what cultural lag conditions the architect to seek out ways of effecting the preindustrial to industrial transformation. There is no easy answer to this question; all one can say at the outset is that it is not really the architect's fault. In large measure this backwardness is the natural result of the institutionalization of the entire building production process within our society.

Systems building is looked upon by its many advocates as essentially a process of reformation: the reorganization of the methods, procedures, and techniques by which a society will produce the built environment. To the systems advocate, construction activity until now has been primitive—a highly individualized, occasionally idiosyncratic process and, most significantly, a process that is unresponsive to a modern society's need for shelter. Systems proponents argue that through a necessary industrialization of the entire construction process, building production will become rationalized and will result in greater economies and production.

Terner and Turner provide a set of clear, coherent, and useful distinctions in the vocabulary of industrialization:

*SYSTEMIZATION*—the design process which gives rise to an assemblage of standardized and correlated construction components used to form a dwelling unit, whereby ad hoc construction methods are deliberately rationalized and regularized, usually in the interests of economy, speed, and quality control.

*INDUSTRIALIZATION*—the process often referred to as "mass production," whereby products, which traditionally may have been made in a hand-crafted and individualized way, are manufactured in larger quantities by a new set of processes which usually imply: 1) a standardization of the final product; 2) a specialization of labor; 3) a concentration of production, purchasing, and marketing; and 4) mechanization or automation of production processes. In addition to increased output, industrialization often implies lower costs because of economies of scale and productivity increases, and uniform and predictable (if not improved) quality of output.

*STANDARDIZATION*—the process whereby parts or products are manufactured in such a way as to be similar enough to be interchangeable, and so that they compare within an accepted or established range of values for size, shape, weight, quality, strength, etc.

*COMPONENTIZATION*—the differentiation of the construction process into relatively separate or autonomous structural or functional subsystems. When these subsystems or sets of components are correlated, as in a housing system, each of the components can be modified, either without substantial changes to the others, or in such a way that a change may be predictably traced through the entire system. In this way one may understand how the initial change affects the series of companion subsystems.

*PREFABRICATION*—the advance production of standardized components or sections of buildings, ready for quick assembly and erection at a building site. Often this production is undertaken at a factory or work area away from the actual site itself.[1]

**History.** New techniques and technologies have always fascinated the architect. Some observers have used the innovative adaptation of new materials and techniques as the lens through which the history of modern architecture can be viewed. Iron and steel were, of course, fascinating to nineteenth-century architects and Sir Joseph Paxton's Crystal Palace might be viewed as a prototype of industrialized construction (Fig. 28). It was built in 1851 for the Great Exhibition in London, dismantled and later put together again in Sydenham in 1854.[2]

The interest in technique was wider than a simple urge to innovate on the part of the architect. There was a heightened consciousness of the need for housing and a growing disillusion with conventional methods of generating an adequate supply of same. This combination of social problem-solving with innovations in material and technique led to Le Corbusier's early attempt at a housing system: The "Dom-ino" (1914). Reinforced concrete was then becoming significant, and the later devastation of the war was to demand some housing response. His system was suggested for postwar reconstruction in Flanders.[3] The Dom-ino system was to provide a minimal shell within which the occupant could fit out to meet specific activities and needs (Fig. 29).[4] During the 1920s, the idea was further developed in the Citrohan project (Fig. 30) and finally realized in Stuttgart in 1927 at the Weissenhof Exhibition organized by the German Werkbund (Fig. 31).

During the period before World War II, there was also a good deal of work done by American architects on prefabrication and industrialization. Two of the better known examples were Buckminster Fuller's Dymaxion House (Fig. 32) and Frank Lloyd Wright's precast concrete panels (Fig. 33).

**Architecture and Systems Today.** While architects like Fuller and Wright provided significant innovations, it was World War II that provided the critical impetus to industrialization of architecture. The war devastation in Europe demanded an immediate response, most obviously in housing but also in other areas, and this was the stimulus to significant experimentation and development of the systems building idea.[5]

Among the milestones in this experimentation was the CLASP school system developed in 1948 by a building consortium for the Hertfordshire County Council in England. Simultaneously, there was in England a broad-scale effort toward the modular coordination of construction. In France, a notable early housing system was devised by the engineer R. Camous. It received substantial support from the French government and resulted initially in a 4,000 unit housing development.

This experimentation and governmental support in Europe led to today's significant diversity in types of building systems there. Essentially there are three generic classifications:

1. *Panel systems* in which whole or partial wall and floor panels function both structurally and as the space enclosing device (Fig. 34).
2. *Skeletal or Frame systems* in which the enclosure and support or structural functions are separated with a standardized skeletal framework functioning in the latter capacity. To this framework, various forms of panel subsystems are then attached to provide spatial enclosure (Figs. 35–37).
3. *Cellular or box systems* in which volumetric modules of room-size boxes or cells are stacked in various ways to both enclose space and support the aggregate structure (Figs. 38, 39).[6] This type of system is perhaps the best known because of Moshe Safdie's Habitat in Montreal.

All three types of systems are further subdivided into heavy and light systems within each category; the heavy systems normally employ concrete or masonry with the lighter systems using plastics, wood, and steel.

While many of these systems are now being introduced into the United States, the American experience generally has been much more limited and much less successful than Europe's.[7] Here, the builder's journal *House and Home*—something of a reality-test for architectural and academic proponents of industrialized housing—sees only four "real" kinds of industrialized housing.

*Single-wide mobile homes, or simply mobiles, are a form of industrialized housing.* They are built on a steel chassis of light wood framing and aluminum skin, are 10′, 12′ or 14′ wide and from 50′ to 70′ long. They do not meet conventional building codes.

*Double-wide mobile homes, or simply double-wides, are a form of industrialized housing.* They are built of two mobile-home sections side by side and are 20′, 24′ and 28′ wide and from 50′ to 70′ long. They may be put on pads like mobiles or on permanent foundations. Many mobile manufacturers call their double-wides sectional homes, but this is misleading. Many double-wides are more heavily constructed than mobiles. But few meet typical building codes, which they would have to do to properly be called sectionals.

*Multifamily modulars, or simply modulars, are a form of industrialized housing.* Buildings are made by assembling a variety of modular boxes into one-, two-, and sometimes three-story structures. Construction materials and specifications are identical to those of conventional low-rise multifamily housing.

*Single-family sectionals, or simply sectionals, are a form of indus-trialized housing.* Like the double-wides, they are made up of two 10', 12' or 14' wide sections attached side by side; unlike double-wides, they meet conventional building codes and are always built on conventional foundations.[8]

In contrast to the larger and heavier European systems, the American experience thus emphasizes a modular or component construction in addition to the low-rise and single-family dwelling unit. The compo-nent idea is illustrated in Figure 39 which shows how the various sys-tem elements can be combined in a given structure; typical of such ele-ments now being prefabricated for component integration are bathroom units (Fig. 40). This idea of componentization leads to another useful semantic distinction in systems-building types: the open versus closed system. The latter is one within which only the elements of a par-ticular manufacturer (or preintegrated design) can be used as opposed to the open system which allows for the use of various manufacturers' products as component units. Architects, ideally, would like to develop the ultimate open system. However, standardization and componentiza-tion are made easier both conceptually and in manufacturing by the use of closed systems (See Figs. 41, 42).

Either because of its intellectual breadth of vision or the unavoidable consequences of the last world war, the European perception of sys-tems building is obviously much more holistic in placing the housing production process within a broad social and economic context; it emphasizes—if compared to the American context—a radical rear-rangement of institutional mechanisms and roles, including those of the architect, builder, manufacturer, material supplier, and financier. The American view, as reflected in the *House and Home* perspective, is more focused on the product itself—the physical entity—with the im-plicit premise that the present institutionalization of the construction process in American society is adequate. These contrasting images are reflected in apparently trivial but ultimately significant definitions of industrialized housing:

> *Systems building:* The application of the systems approach to con-struction, normally resulting in the organization of programming, planning, design, financing, manufacturing, construction, and evalua-tion of buildings under single, or highly coordinated, management into an efficient total process.[9]

> *Building system:* An assembly of building subsystems and com-ponents, and the rules for putting them together in a building. Nor-mally these components are mass-produced and used for specific generic projects in a construction program.[10]

Thus, while the American focus has been on the physical entity, as opposed to the larger, more holistic process, of systems building, various forces have combined to broaden our view in favor of this larger context. Significant among them are the sheer magnitude of housing demand in the United States, the relatively recent introduction here of the large-scale European building systems with a consequent introduction of the management and organizational techniques which accompany the systems, the evaluation of the rather sad experience of Operation Breakthrough. The somewhat volatile nature of building systems and modular manufacturers' stocks on the American market with consequent Wall Street reaction and financing difficulties has not helped. To predict, then, what industrialization might become in the United States requires our relating the American experience to the larger European holistic vision and that leads, essentially, to sheer speculation. In this connection, perhaps the best summary definition of American industrialized building—with implications as to what is at least conceptually possible—is the model of construction products developed by Miriam S. Eldar of Sweets Information Services. Her Construction Matrix depicts a model of the entire built environment with a hierarchical breakdown of its constituents: explicit products and implicit processes (Fig. 43). The terms she uses in defining her perception of the construction world are:

*BASIC PRODUCT TYPE*—the classification of products into six types by structural and functional complexity—Basic Material, Unit, Assembly, System, Module, Facility.

*BASIC MATERIAL*—a simple product, pre-formed or formed-in-place, adaptable to a variety of uses, whether in the manufacture of construction products or applied directly into the fabric of conventional construction.
Examples—sheet glass; brick; resilient flooring; sealant, etc.

*UNIT*—a built element complete in itself which can be used independently, or become a component of a larger whole (Assembly, System, etc.). No assembly occurs at site, only installation.
Examples—pre-hung door and frame—can be used alone, or as a component in an enclosure assembly; kitchen range—can become a component in an assembly of cabinets and appliances.

*ASSEMBLY*—a pre-designed and pre-fitted whole, comprised of a number of components with a high degree of interchangeability. Three subtypes can be recognized:

*Built Assembly*—components, when fitted together, form a built element.
Examples—curtain wall; hung ceiling, etc.

*Network*—components, when fitted together, do not form a single built element.
Example—electrical distribution network.

*Coordinated Group*—group of autonomous units, unified by and/all of the following—sizing, function, shaping, materials, etc.
Examples—group of furniture; group of door hardware.

SYSTEM—built element(s), comprised of a number of components which, when fitted together, integrate at least *three* of the basic building functions and have a highly predictable performance as a whole.
Examples—integrated ceiling, structure-enclosure-environmental control system.

MODULE—a large built element which integrates at least three of the basic building functions, is self-supporting and in addition encloses space.
Examples—Dwelling modules for multiple-dwelling facilities; complete kitchen; bathroom; service-core module; clean room; audiometric room.

FACILITY—a built environment which satisfies all building functions to the extent required in order to serve one or more human needs.
Examples—single-family dwelling; greenhouse complete with environmental control.[11]

**The Argument for Systems Building.** In the historical development of the systems building idea, the analogy most often employed in support of industrialization was, of course, the automobile. After all, the reasoning went, if Henry Ford could turn out a complex system—the Model T—at a rate of X million a year and bring down the unit cost through economies of scale, why not adopt a similar technique (i.e., mass or serial production) and realize similar economies of scale in producing housing?

This shift of the construction process from the building site to the factory ideally brings many advantages: lower production costs, greater cost control, less time spent in the production process, economies of scale, greater savings both in the utilization of labor and through the employment of cheaper industrial labor rather than the crafts-union type of on-site labor traditionally employed (essentially a lessening of the skills required to produce housing), improved quality control; and better management achieved by the use of advanced management techniques not normally associated with conventional construction work. To these accrue peripheral benefits such as faster production and a freeing of the productive process from the vagaries of weather which obviously limits on-site work.[12]

Regardless of whether one takes the more holistic, academic, or theoretical view or the day-to-day, pragmatic builder's perspective, the arguments for industrialization are substantial and, in fact, are reflected in actual production figures. For one thing, industrialized or systems building does take less time to produce. Absolute figures are often bantered back and forth and consequently there is no definitive ratio of industrialized production time vis-a-vis conventional or "stick-built" construction but specific examples are illuminating. Robert E. Platts, for instance, in assessing the performance of a particular European building system, notes:

> This isn't luxury housing or somebody's stunt. This is high-volume people's housing, and very high quality work. It's yielding quality apartments in under 500 man-hours, including factory and field, as against the 1,100 man-hours we take for lesser quality apartments in North America.[13]

Savings are realized not just through the reduction in absolute hours of labor, there is also a significant substitution in the cost of that labor. This substitution of industrialized skills for craft skills can be enormously significant. Whereas the conventional contractor pays carpenters, plumbers, and electricians $8 per hour and up, the modular builder employs relatively unskilled labor at an average of just above $3.00 an hour.[14] While it is difficult to get reliable figures on the American experience, the general data on trade performance adds additional support to the economy argument for industrialized production. For example, the trend in hourly earnings of industrial versus construction workers (Fig. 44) indicates at least the maintenance if not an actual growth of a significant wage differential. Moreover, the trend in productivity of the construction worker versus his industrial counterpart is in favor of the latter (Fig. 45).

Finally, when one compares the resultant trend in actual production of conventional versus manufactured units, the gain in the latter is substantial. Here, of course, mobile homes are included as a form of industrialized housing and the trend is up (Fig. 46). Moreover, when one compares this with mobile homes as a percentage of conventional one-family starts the trend is again toward industrialization. What is signficant in these patterns is that it is the smaller modular unit that is successful in the United States as opposed to the European experience.

**The Argument against Systems Building.** A rather pernicious argument against industrialized building has been voiced by the architectural profession itself. Alarmed by standardization and its perceived consequences—the relative reduction of the importance of architect vis-à-vis the manufacturer, supplier, engineer, and even the salesman—the design professions have protested industrialization's "constraints on creativity." After all, the reasoning went this time, didn't Henry Ford also have a dictum which made explicit the inevitable design constraints implicit in the process of standardization, "give them any color they want as long as it's black." Where was the architect's freedom to express himself?

There is a great deal of diversity of opinion among architects on this point. While many have tacitly avoided the issue of industrialization, perhaps feeling that the limitation on design freedom was significant, other architects have welcomed it: among the latter Gropius and Le Corbusier. The architectural advocates have found greater opportunities for creativity in industrialization as the architect moved up on the scale at which he designs the built environment: from directing the cutting and nailing of pieces of wood, to the assembly of modules—packages within which others have already done the cutting and nailing.

A stronger argument against industrialization is that existing "stick-built" methods are not all that bad. In fact, there is a question as to whether the United States really has experienced any substantial industrialization of the housing-production process. Some argue that industrialized housing is essentially a congeries of on-site procedures and techniques that happen to have moved indoors; the true potential in the substitution of mechanization in the factory for craft labor on the site is yet to be realized.

Building codes are also a significant restraint on industrialization and, while the codes are restrictive per se, the fact that they vary from one locality to another causes even greater difficulty. The producer of manufactured housing has to be assured of the legal acceptability of his product at all locations in order to get a reasonable portion of the housing market. Codes, however, provide pressure groups which are a powerful resistant to modular and prefabricated housing in their localities. Fundamentally, however, it is the lack of jurisdictional consistency between state and local codes that inhibits industrialization.

Labor unions have been a major obstacle in the rationalization of housing construction since it threatens existing jobs and provides a mechanism for the substitution of cheaper industrial labor for a more expensive crafts-union, on-site construction crew as we have seen. This reluctance of the unions to permit prefabrication—and ultimately industrialization—was illustrated notoriously and well in the "Philadelphia Door" case of the late 1960s.[15] The issue arose when a contractor attempted to use some thirty-six hundred prefabricated doors on a large project. These doors were premortised and fitted; ready for hanging. The carpenters refused to install the doors, pointing out a provision in their contract which prohibited such prefabrication. Consequently, the contractor returned the doors and received blank ones upon which the carpenters were allowed to do the traditional mortising and fitting work. The issue was ultimately taken to the Supreme Court as a possible violation of the Landrum-Griffin Act which prohibits an employer from refraining to handle the products of any other employer. The union won; the Court ruled the Act was not violated and the union's purpose was only to prevent its members from losing work due to the off-site prefabrication of work that was previously done on-site. Similar issues have arisen about various other components of housing such as prefabricated roof trusses, preplumbed wet-walls, cabinet work, and other elements that can be assembled as complete components.

The unions are not alone in a preference for traditional methods. Building suppliers still want traditional routes to be followed in the distribution of materials via their outlets to the ultimate consumer. Industrialization of housing production demands that there be a direct and cohesive relationship between materials manufacturers and the factory operation. This obviously affects the interests of the local distributors of building materials and they usually resist. There is also a traditional lack of adherence to rigorous schedules in the deliveries of materials to construction sites whereas such a rigor is necessary for a factory operation in which the production line must have deliveries on a very tight schedule in order to exploit fully the productive capacity of both its machines and men. The supplier-factory relationship has not been made any easier in the United States by the many financial failures of industrialized housing firms; suppliers are now skeptical about dealing with such manufacturers.

Transportation is also important. Beyond a set distance from a plant, normally 300 miles, the modular or manufactured unit becomes increasingly less competitive against conventional "stick-built" housing. The rough measuring unit of transportation cost at present is approximately a dollar per trailer load per mile. Most modular houses today comprise at least two trailer loads and trucking is still the basic mode of transportation although various other means of transporting modular units have been tried, including rail, helicopter, and barges.

While unions, suppliers, and transportation have all provided significant obstacles to the industrialization of housing in the U.S.A., perhaps the most significant drawback lies in the very organization of the housing market itself. The stumbling block is the aggregation and continuity of the market—meaning essentially that if sufficient capital is to be invested in plant and the other aspects of industrialization, there must be a return on this capital which will justify the expenditure. This, in turn, demands an appropriate level of housing demand and moreover a consistency in this level over time.

To return to the traditional automobile analogy, Ford or General Motors are assured under normal circumstances of a relatively stable and predictable demand for their product. This allows them to plan for capital expenditure, the employment of labor, and the other necessary conditions of factory production all premised upon a relatively stable and, in most instances, continually increasing demand. Today, there is simply not an aggregation of the housing market at a level sufficient to provide an adequate return on the high capital investment of the industrialized houser. What we have in the United States is a complex of local, highly individualized, and relatively small markets. Moreover, seasonal and other cyclic variations have augured against any coherent system of predictability in the market. In a different context, this point has been stressed by Gunnar Myrdal in his observation that the United States government uses the construction industry as an instrument for either heating up or cooling down the economy.[16] Some of the other arguments against industrialization are summarized by Alberto F. Trevino:

> Supporting subsystems which have to be in place and operational to allow a house to be manufactured. . . .
>
> FINANCING ALTERNATIVES—New methods of construction payment for housing have to be developed. Most lending institutions are accustomed to progress payments which are now unnecessary with manufacturing. Problems also include guarantee of payment to the manufacturer, separation of site and housing payments and transportation payments.
>
> TRANSPORTATION & DELIVERY—Adequate carriers and fair regulations by the Public Utility Commission; a developed highway system with adequate height clearance which allows carriers to transport units efficiently to sites.
>
> CONSTRUCTION & ERECTION—Sufficient machinery and skilled manpower to set large numbers of units per day. If heavy box systems are used, are there sufficient large cranes in the region that are available to set hundreds of units per day?
>
> APPROVALS & LEGISLATION—In-plant inspection systems to coordinate with site development inspections and final inspection and testing. State housing legislation such as was passed recently in California (Assembly Bill 1971).
>
> RAW MATERIAL—Available material in quantity to satisfy all markets and development of new materials to stand the stress of shipment.
>
> INSURANCE—Adequate methods to protect the manufacturer from liability or loss once the unit leaves the manufacturing facility.
>
> LABOR—Training of labor within the existing construction unions to meet the demands of the future and to integrate the various trades into a productive unit.
>
> CONSUMER TESTING—Manufacturing techniques will spawn new forms which must be market tested.
>
> COMMUNITY DEVELOPMENT—Land planning alternatives which program a different sequence of events; possibly allowing landscaping to precede final erection.[17]

**Summary.** Historically, the creative architect has taken certain elements of advancing technology—both material and techniques—and has incorporated them in new architectural expressions. Some of the great American innovators were in this tradition: Bogardus, Sullivan, Wright. In systems building, however, the architect is not simply incorporating new technology; he is asking society to transform radically its economic organization so as to provide shelter more efficiently. The various issues involved: aggregation of the market, response of the unions, and so forth, are more than simple questions of technique; they generate fundamental economic, social, and political issues that are not normally the architect's prerogative to resolve. In any event, the mandate to intervene so radically in this social transformation is not yet manifest. While we might be amused by the "architect as ultimate social arbiter" (dicta in the manifestos of Wright, Le Corbusier, and others), realism forces us to recognize that it is one thing to be radical in the use of material and technique, it is quite another to attempt radical transformations of society.

Presumptuousness has also been an impediment to the development of industrialized housing in another sense. Architects have generally favored relatively dense, high-rise housing in an essentially urban milieu as the appropriate context for development of the systems building idea. Yet the marked preference of the American people—as they vote in the market place—is for the single-family, detached dwelling on its own lot. This contrast in value systems has caused the technically competent (i.e., architects and allied arts) to place the systems building idea in one format while the housing consumer and consequently the market have chosen an entirely different expression.

Political naiveté has also been a drawback. Systems advocates do not seem to realize that decision-making in a democratic society is essentially a process of bargaining. The costs of rationalizing the industry along the lines they visualize would mean a direct threat and/or substantial change to unions, mortgage bankers, building manufacturers, producers, distributors, suppliers, and others. Opposition arises and consequently sentiment must be expressed in the political arena which, minimally, will equate the demand for rationalizing the building industry with the costs to be born by present actors within the existing institutional structure. This kind of collaboration does not seem likely in the foreseeable future; hence, the performance of systems building in the United States has been rather marginal despite the inherent elegance of its logic.

Despite its advocates' lack of sensitivity to social and political realities, the application of the systems idea to systems-building has been one of the former's most fruitful expressions. While both rationality and holism are not as self-consciously attempted here as in other expressions of the systems idea, both are manifest in significant ways. The war in Europe and the general industrial milieu within which thinking about systems building was formulated were as significant in the development of the idea as was self-conscious intellection as to how housing should be produced by a society. The lesson would appear to be that while logic or rationality may be theoretically persuasive, social momentum must be generated if the systems idea is to become effective.

## 9. The Systems Idea in Systemizing the Planning and Design Process

The most ambitious undertaking by systems advocates has been their effort to bring rational coherence into planning and architectural design. After industrialization of the physical artifact, it is perhaps inevitable that one begins to industrialize one's thought processes; in moving beyond the preindustrial crafts method of production we also move beyond traditional design methodology.

Work in this area is a relatively recent phenomenon having gained momentum only in the late 1950s and 60s. The principal incentive was dissatisfaction with traditional methodologies in the face of increasingly complex problems. More to the point, it was a dissatisfaction about the way design and planning methods were taught in the schools, as evidenced by the fact that most of the advocates of this rationalization have been academics.[1]

One of the strongest incentives to rationalization was the demonstrated success of the systems approach in the design of aerospace and weapons systems. The accomplishments of designers in these areas were manifest and impressive; within a relatively short time span, they programmed, planned, designed, and built enormously complex physical systems that did very interesting things—like fly us to the moon.

In contrast to the shortcomings—real and imagined—of traditional architectural and city planning methods, the weapons systems process was both well-defined and rigorously logical: as one progressed over time, certain benchmarks were established and rigidly adhered to. These benchmarks took the form of so-called "baselines" which were essentially definitions of agreed-upon conditions and intentions at critical moments in time, including those times at which the entire design effort was to be reviewed. The normal sequence was as follows:

1. Concept formulation leading to definition of the program-requirements baseline.
2. Program changes leading ultimately to a design-requirements baseline.
3. Modification of design-requirements as the design itself progressed, with the ultimate incorporation of design changes in the form of a final product-configuration baseline (a detailed definition of the systems' physical and performance characteristics).

The system itself was built from this last phase of the situation, and while modifications could take place in the physical systems after they were in use, such changes were intended to be essentially trivial in nature. The best thing that can be said about this process is that it worked. Diverse skills and capacities were organized in an enormously complex division of labor, and the systems developed performed close to their prior specification. Critics will argue that given the resources with which the planners and designers worked, success was insured. There were, in fact, major fiascos, perhaps the most prominent being the TFX-F-111 decision. But inquiry into this disaster indicates that there was a lack of clear definition of what was intended at the outset rather than inherent defects in the analytic process itself.[2]

The overall use of the approach in developing weapons systems, however, was largely successful and as a consequence architectural and planning methodologies were critically scrutinized in the light of its imperatives. There were some rather facile transfers of the design technology to architectural and urban physical systems.[3] Here, however, the work was not with the relatively simple and deterministic performance of hardware systems; it was enmeshed in the historically difficult task of attempting to define what is "good" or desirable in architecture and cities and how one goes about realizing these conditions [4] as we have discussed in Chapter 5.

As these methodologies developed in architecture and planning, one of the more interesting and logical rationalizations was to go directly to the root of the problem and attempt to determine so-called "user needs": to identify specific user populations of buildings and cities, to specify the values and cultural conditions existing among these populations and, finally, what the consequent but varying requirements of use might be. Using the insights of the behavioral sciences, one then attempted to formulate an operationally useful definition of user needs.[5]

The danger here is of putting human beings into tight sets of categories of both observed and probable future behaviors in order to specify "need" on the basis of such categories. That process allows for little human mobility in a geographic, social, or cultural sense. For example, a certain cultural or ethnic category would probably conjure up in the mind of the designer a specific type of environment appropriate to the behavior associated with the group. Yet, if one does not allow for some form of categoric disaggregations of urban populations, little can be said, in an operational sense, about what users who are culturally distinct might need. This forces one back into traditional architectural ideologies (or delusions or insights, depending on one's perception) like those which Frank Lloyd Wright or Le Corbusier would have asserted as to what people should have. Some different attitudes about the identification of these varying "user needs" are shown in Figure 47.

Another difficulty in identifying user needs arises in the prior definition of user categories; this is not as easy as it might first appear to be. With respect to housing, for example, one instinctively thinks of the group occupying a house as the "user" in an almost exclusive sense. Theorists in this area have developed larger perspectives of neighborhoods, sectors, and city users, but, even if one does define some sort of hierarchy of use in the ultimate composition of the city, as one extrapolates from the basic unit of the house as conditioned by the relevant economic and other parameters, will this be sufficient for a determination of a housing program requirements baseline? [6] Various other factors have also been institutionalized in the housing design and production process (Fig. 48). All the actors cited here have their own visions as to how so-called "user needs" should be defined. Indeed, many have seen to it that their particular perception of need has become institutionalized in either law or administrative regulation.

Because of these difficulties, it seems that any meaningful specification of user need and consequent design prescriptions must be viewed in the context of some value system. Value systems establish the contexts within which specific criteria emerge as significant or trivial. This then allows needs to be ordered by distinctions of preferredness of criteria of utility (Fig. 49).

We will recall from Part I that definitions of the planning and design process are essentially elaborations on models of choice. Fundamentally, the planner and decision-maker wants to know what to choose when confronted with certain options and the probable consequences of those options (see Fig. 50). At the architectural scale this usually works out to a model resembling that depicted in Figure 51 or (with greater attention given to the varying value systems of potential users) to that shown in Figure 52. At the urban-planning scale the social, political, and institutional constraints become more numerous with consequent implications for model complexity (see Figs. 53 and 54).

To place these models in a useful perspective, one must comment on what might be called the sociology of the "methods movement." Largely developed in the last decade, it is essentially the work of academics: younger professors and students. Practicing professionals are conspiciously absent from its ranks. In this context, doubts arise occasionally about the potential "real world" applicability of some of the techniques. With the exception of those areas of planning which were engineering oriented (e.g., transportation, water supply), real world use of the newer methods has been largely restricted to the adaptation of basic management techniques whose utility was demonstrated in military and space hardware design, e.g., network models such as CPM (Critical Path Method) and PERT (Program Evaluation and Review Technique) which are concerned with the scheduling of tasks and the anticipation of their interrelatedness, and of the probable resources they might require.[7] There has also been substantial use of financial and accounting models and management techniques adapted from the

72

business world and, to a lesser extent, the use of computer graphics as a design tool.

In urban planning, there is now substantial use of various statistics packages, time series, and forecasting models and—increasingly— simulation models. Where there was an easy transfer of technique because of the general applicability of the problem-solving method, certain programs have also been adapted, for example the CRAFT [8] and other area allocation programs where the parameters and their relationships can be well defined as in the design of a factory. There has also been an ongoing use of engineering economy techniques for the assessment of probable financial consequences in project analysis. Thus if one were to characterize the techniques that have been adopted to date, they are essentially those of a relatively objective nature useful in dealing with a quantifiable reality as opposed to those that might be concerned with the more fundamental and difficult problems of social, psychosocial, and aesthetic determinants of design. In a sense then, these techniques can be seen as extensions of an engineering technology which has always underlain architectural design: just as one would hesitate to attempt structural analysis by intuitive methods alone, we now have exacting methods for other tasks which will eliminate a reliance on intuition.

These more exacting methods must be recognized, however, as somewhat pedestrian in the sense that most of the theoretical postulates which deal with the softer and more difficult areas of planning and design, do so with a corresponding lessening of real world application and testing. In sum, one can merely observe that systems constitute a rapidly developing area and one in which the benefits are yet to be fully realized.[9] While there will no doubt be a significant discard of some of the present attempts to rationalize the planning and design process, the attempt itself is essential. Moreover, while relatively sophisticated partial models of the process exist, the ultimate integration of these within an overall model of how one proceeds in planning and design is still a distant goal.

# III
# Utility
# of the
# Idea

## 10.  A General Assessment of the Systems Approach in Architecture and Urban Planning

Of possible alternative formats by which to examine the utility of the systems idea, the most useful is directed at the idea's central themes of holism and rationality. We can begin with a critique of the latter since it is ultimately the more cogent and revealing.

Rationality, in the current use of the systems idea in areas like defense analysis and, to a lesser extent, by operations analysts generally, is largely economics thought. Consequently, the abstraction and model-building process is associated with an understanding of variables and an ultimate definition of "system" focused upon that which is economic in nature—with the implicit exclusion of other manifestations of reality. Given this focus, one looks largely for efficiency in the use of resources:

> The basic theme which runs through economic or policy analysis is that of economic efficiency, broadly defined. Considered most simply, economic efficiency is achieved when the value of what is produced by any set of resources exceeds by as much as possible the value of the resources used; or when the least valuable set of resources is utilized in producing any particular worthwhile output. Both benefit-cost and cost-effectiveness analyses apply this economic efficiency principle to public expenditure choices.[1]

Traditionally such use of resources has been calculated by a fictional character known as "economic man": an idealized abstraction of an actor who operated with perfect knowledge of available options and their relative costs and benefits. Equipped with a coherent perception

**74**

of his own utility, this hypothetical and calculating character was completely rational (i.e., unemotional, nonintuitive) in his decisions. While such a construct is purely imaginary, its use was sufficiently valid to allow economists to develop theories which were later confirmed by reasonably accurate measurements of market activity.

As we have discussed at length, this success fostered the image of economic rationality as a sufficient guide in all decision-making, even that which was social in nature—wherein groups such as building committees, communities, and planning commissions were to make choices about the use of resources in pursuit of goals. Paul Diesing says:

> Economic rationality has long seemed to many people to be the very essence of all rational planning and decision-making. To many it has seemed self-evident that the way to make reasonable decisions is to evaluate ends in terms of alternative costs and to allocate means in terms of marginal (or comparative) utility. The ultimate objective of all decisions and actions has been assumed to be the maximization of welfare or satisfaction or output per man hour or some other measurable quantity. The principles and rules for this kind of decision-making have been developed in detail by economists, justified by various kinds of utilitarian philosophers, and embodied in countless policy proposals and discussions.

> In recent decades the defects of utilitarianism and the limitations of the economic viewpoint have become apparent. Yet the economic model of decision-making has retained its appeal, in spite of its limitations, because of its mathematical clarity and simplicity and because of the lack of any clear alternative planning method. Recognition of its limitations has led to several kinds of modification, but these have not changed its basic character.[2]

The clarity and simplicity of such reasoning demands that the objectives or ends sought be defined with a rigor and specificity that would make them operational in the decision-making process.[3] This normally means at least a minimal amount of quantification. But is this ever possible in complex problem-solving without a gross distortion of reality? As E. S. Quade observes, in decision situations:

> Objectives are not, in fact, agreed upon. The choice, while ostensibly between alternatives, is really between objectives or ends and non-analytic methods must be used for a final reconciliation of views.[4]

Such a reconciliation is necessary in city planning because the rigorous spelling out of objectives would generate significant tensions within normal political processes. It directly contradicts the interests of political decision-makers who, more often than not, would prefer objectives that remain rather hazy and vague. Once the objectives are defined with any degree of specificity, the rational voter can presumably identify the terms of trade-off that have been established between his own personal utility ("preferredness" or interests) and that stated by his political representative to be in the public interest.

One effect of this would be to draw uncomfortable attention to pork-barrel politics. A voter who is informed of assumed objectives and the terms of trade-off between them, would be equipped to assess a decision process which allocated only ten million dollars to a HUD housing program while in the same fiscal year allowed a hundred million dollars to be spent by the Department of the Interior to build a relatively useless dam somewhere. Alternatively, the normally tenuous coalition and vote-trading mechanisms of a city council where Councilman A will vote for Bill X in order to get Councilman B's support of Bill Y is simply too subtle to be properly apprehended by and made amenable to systems analytic techniques.

In such situations a rigorous specifying of objectives and the more general and pervasive objectivity of the systems analytic process itself might be seen to be incongruent with democratic political mechanisms in which a certain ambiguity is useful in maintaining minimal social cohesion within a milieu of imperfectly sublimated tension; a milieu in which the political actor wants to appear to be giving all things to all constituencies. The analytic process would force greater political contentiousness as problem situations become more clearly defined and objectives more specific. It is not an unreasonable conjecture to foresee the possibility that systems analysis could, in fact, change the political process itself, at least to the extent that planning could become more significant as a mode of social choice with the attendant diminution of "politics as usual."

The political dimension aside, how well *can* ends and objectives be specified? Charles Hitch says that

> We must learn to look at objectives as critically and as professionally as we look at our models and our other inputs. We may, of course, begin with tentative objectives, but we must expect to modify or replace them as we learn about the systems we are studying—and related systems. The feedback on objectives may in some cases be the most important result of our study. We have never undertaken a major system study at RAND in which we are able to define satisfactory objectives at the beginning of the study.[5]

It must be remembered that this is a common experience in architecture. In initiating a project, the architect normally has some sort of brief or definition of intent and resources from the client. As the work proceeds, the consequences of these client-specified objectives are determined by the architect. This process of clarification inevitably results in a rethinking of priorities by the client in the context of the a priori resource constraints, the implications of which become more explicit and significant as the work progresses. It is the common experience of architects and their clients to go through a period of mutual consternation as the client recognizes the impossibility of achieving his initial intent with the available resources. This issue becomes particularly acute when the client is a building committee or the board of, say, a school—the various members of which tacitly assumed coalition ends without recognizing that their personal perception of ends may be in conflict. As the implications of resource constraints become better defined, there may be a falling out among members of the board with different value systems. Board member A might value the utilitarian function of the building; board member B, its aesthetic quality; board member C, its symbolic or community service aspect, and so forth. Thus, the simple assumption of consensus as to those ends which can be identified at the outset and pursued without modification through completion of a project—either at the architectural or the city-planning scale—can prove to be a dangerous presumption resulting in a great deal of difficulty in the later design process.

While it may be useful economic reasoning to see each individual as a utility maximizer who knows what he wants and reaches decisions after a careful cost-reward assessment of available options, is this a reasonable model for collective decisions? The difficulties here are classic: how, for example, does one make interpersonal utility comparisons? If A likes dense, high-rise inner-city living and B likes exurban, bucolic, pastoral life, how does planner C get into the "preferredness" assessment mechanism within the minds of A and B so as to determine how much A will be pained by living B's way and vice versa? Unless one has a market within which preferences or measures of preferredness can be expressed (or alternately, some form of market surrogate), it is virtually impossible to resolve this issue.

Presuming that such an issue were in fact resolved, how does one then aggregate the various measures of individual preference into some overall social "welfare function"? Abram Bergson defines a welfare function as:

> W, the value of which is understood to depend on all the variables that might be considered as affecting welfare: the amounts of each and every kind of goods consumed by and service performed by each and every household, the amount of each and every kind of capital investment undertaken, and so on. The welfare function is understood initially to be entirely general in character; its shape is determined by the specific *decisions on ends* that are introduced into the analysis. Given the decisions on ends, the welfare function is transformed into a scale of values for the evaluation of alternative uses of resources.[6]

Again, it is precisely these decisions on ends that present the problem. To assume that there are overall ends to which we all subscribe and, in turn, within which these measures of preferredness can be subsumed is naive. It is also naive in the sense that one cannot really ascertain what is "cost" and what is "benefit" in the solution of a social problem without a fixed definition of ends. In city planning, for example, the costs and benefits will be differentially distributed both in a geographic sense and with respect to various social groups. Take for example the recurring issue of that Lower Manhattan Expressway in New York City. Why should the Chinese and Italian communities along its Canal Street path incur a direct and significant cost—the possible disintegration of their communities—in order to generate a diffuse, somewhat nebulous but broadly distributed benefit: facilitating the movement of people wishing to go from Long Island to New Jersey and vice versa? How does one assess the magnitude of the cost imposed on relatively few people with the comparatively trivial benefit given millions of highway users? Or, to take a quite different example, one might focus on the issue of equity arising from the costs and benefits distributed to the recipients in the light of their comparative characteristics as social or income groups. If one defines a welfare function, is it not inevitable that there will be implications for income redistribution within it and are we to assume agreement on this as a premise for planning?

In another context, Fred M. Frohock has noted the relativistic consequence of this argument:

> Since an object is good or bad by virtue of subjective response, then the same object can be either good or bad under varying conditions and with different persons encountering it.[7]

Thus, at the scale of city and regional planning, geographic and class differentials combine, in their value implications, to generate significant barriers to the operational use of the systems analytic techniques. As a result, the question of values becomes significant. But there is a difficulty here in that the introduction of values makes questionable the entire "scientific" nature of the systems analytic procedure. At least as far back as Max Weber there is a tradition in social science that the scientist qua scientist cannot suggest what *ought* to be done: he is limited to description and explanation of phenomena as opposed to any suggestion as to what might be done about the situation. Weber says that to introduce norms into scientific analysis is simply to be unscientific.

The planner, however, is caught up in a normative quagmire. In light of the preceding examples, value decisions are obviously the essence of what he is about. If he is socially responsible, he not only describes and assesses existing conditions, but he also recommends policies to the political decision-maker. Here he confronts the dilemma of achieving a necessary objectivity while not really knowing how scientific his analysis is.

Not the least of the problems is to distinguish between fact and value in any given problem situation. The properties we might perceive are derivative of our mental framework—our value systems. This is implicit in our earlier discussion of abstract and model-building: we are conditioned by education, experience, familiarity with a problem, etc., to perceive only *selected* phenomena. Moreover, the function of a model is not to give an exact picture of the real world in the sense that one describes all the elements that might exist in that world, their probable attributes, and all their possible relationships with other elements. On the contrary, the model is an abstraction from reality of only those elements that will have some significance in defining the problem and in forming the solution process. The opportunity for a selective or biased representation of reality in determining this significance is obvious.

The complexity of the system analyzed also affects how one looks at the factual content of a problem situation. In defense systems, a significant amount of the data can be perceived as "hard" or objective fact, particularly so in decisions involving weapons hardware. But in social decisions, a factual base is much more difficult to establish. Even when established, the significance of a given set of facts is quite relative, since it is the use and abuse made of the factual matter within the sociopolitical arena that defines the essential measures of significance for the decision-maker.[8] In effect, one must account for the familiar but inevitably nebulous problems of value, fact, and definitions of the "public interest."

In the evaluation of facts deemed relevant in a given problem situation, Hyman Rickover finds the systems approach wanting, ironically, because it is not holistic enough.[9] He perceives a tendency on the part of analysts to isolate a few potentially irrelevant variables within the overall situation and to focus the analysis on them. Specifically, he contends that analysts tend to look at cost and avoid assessments of effectiveness. The former are always measurable—however arbitrarily—but the latter are much more difficult to determine—the measures available often being few, unreliable, and arguable. Analysis also tends to focus on the economic aspects of the problem; these at least being somehow perceptible. Another weakness in establishing the factual base for analysis is the tendency to emphasize quantifiable variables simply because they are manipulable in mathematical modeling.

All this points up an inherent general weakness of the rational model: the analyst does not have the resources—time, energy, money, knowledge—to consider all relevant options in making a decision. Therefore, in building his model, he abstracts selectively and sparingly from the real world with significant possibility for its distortion.[10]

In attempting to focus upon the economic aspects of a given problem—and, at least implicitly, defining all else as economically derivative—the analyst imposes severe subjective limitations on what is alleged to be an objective process; one economist summed up some architectural concerns here in the title of an article: "Quality: Stepchild of the Economist." C. E. M. Joad, the British philosopher, raises the issue in his "Scientific Analysis of a Bach Fugue," [11] in which he briefly runs through an "objective" analysis of a fugue and delineates the specific conditions that will occur when a person listens to music and assesses it in a value-free context. In the latter, only physical properties are significant: the acoustics, the generation of sound waves, their perception by the ear, their transmission to bone structure, the biological phenomena which occur in the ear, and various other elements of the perceptual apparatus and, ultimately—the transfer of this information to the brain. The objective or scientific analysis of the act—free from values—is really a caricature of what one's appreciation of music might be. Thus the definition of situations as essentially economic in nature to the exclusion of other values—aesthetic, moral, social, political— implies an enormous reductionism. In this context, it is not too presumptuous to conclude that if notable architecture or city design is to be created it cannot be generated by allegedly objective analytic techniques alone.

Consequently, criticism of systems analysis has focused on the too-facile transfer of the approach from defense systems to social systems, seen by some to be unjustified reductionism. The point is that the latter are much more complex than the former and that such a transfer of analytic technique may be gratuitous or unperceptive and, ultimately, unwise.[12] Ida Hoos has explained this critical position:

> There is . . . [an] assumption that since large scale, complex systems have been "managed" by use of certain techniques, then social systems, which are often large and always complex, can be "managed" in like fashion. This presupposes similarity of structure, with social systems reducible to measurable, controllable components, all of whose relationships are fully recognized, appreciated and amenable to manipulation. The very characteristics which distinguish social from other species of systems render them resistant to treatment that tries to force them into analytically tractable shape.
>
> (1) They defy definition as to objective, philosophy and scope. What kind of definition of a welfare system can be regarded as valid—that which encompasses the shortcomings of *other* systems, such as health, education, employment or the one which focuses on individual inadequacy? A definition of welfare as an entity without explicit reference and perturbation of *other* systems of society would be meaningless. Moreover, the definition depends on the point of view and the ideological posture. The system looks very different to the administrator, the recipient, the Black Power advocate, the social critic and the politician.
>
> (2) "Solution" of social problems is never achieved. You do not

"solve" the problems of health or transportation. Consequently, where you start and where you stop is purely arbitrary, and usually a reflection of the amount of money the government has to fund the particular analysis.

(3) Despite the semblance of precision, there are no right or wrong, true or false solutions. Consequently, it is presumptuous to label as wrong anything being done now and right that which looks good on paper. By concentrating on minuscule portions or isolated variables simply because they are quantifiable, the technique may actually lead to results which are irrelevant and inappropriate.[13]

The Hoos criticism, while significant and revealing, is directed largely at the systems idea's "rational content." Similar comments have been aimed at its holism but such criticism has been much less telling. Largely these critics have extracted several elements of the preceding —the limited intellectual and resource capacities which might be directed at a problem—and have tried to show how these constraints limit one's ability to develop a holistic view. The critique is based on the fact that one can never really approximate the comprehensive ideal of a thorough and rigorous examination of *all* the options leading to the achievement of a defined objective, assuming that the objective itself can first be defined with sufficient clarity to perceive the options. At issue is the degree of holism exhibited by a city or other geopolitical unit:

> Comprehensive planning, as planners view it, is so often a version of the impossible dream, that the ideal political requirements for it seldom exist in this society. For example, comprehensive planning seems to require a relatively high degree of centralization, so that rational, consistent and efficient management and coordination of all the inputs and outputs related to the planning process might result. Yet the actual system of democratic government stresses the dispersion of power decisions by the many rather than by the few. It spawns special interests, for it is not always characterized by a dominant or visible common purpose. It is not always a system amenable to central coordination. Whatever ideological refuge the serious planner can find in support of comprehensive planning and policy-making, it is insufficient to accommodate the practical problems of making comprehensiveness work.[14]

An insufficient critique, no doubt, but the effort is essential. From the time of Plato many have made the effort to achieve holistic vision, however short of the ideal the ultimate outcome might be.

While in this chapter, we have summarized the more significant shortcomings of the systems approach, it is by no means an exhaustive critique. For that, one might look to Braybrooke and Lindblom who found the approach wanting because of:

1. man's limited intellectual capacities
2. his limited knowledge
3. the costliness of analysis
4. the analyst's inevitable failure to construct a complete rational-deductive system or welfare function
5. interdependencies between fact and value
6. the openness of the systems to be analyzed
7. the analyst's need for strategic sequences to guide analysis and evaluation
8. the diversity of forms in which policy problems actually arise [15]

More recently, Douglas Lee assessed the use of large-scale models in urban planning and, by implication, the systems approach. In addition to several of the preceding criticisms, he noted that

(1) the models were designed to replicate too complex a system in a single shot, and (2) they were expected to serve too many purposes at the same time

in addition to being excessively hungry for data, too gross with respect to the level of detail required in policy-making, and too expensive and complicated for operational use.[16]

## 11. Some Counter-Models and Their Utility

**Naive Alternatives.** Useful criticism is normally given in terms of alternatives and, while the previous remarks may seem severe at times, it is in relation to potential alternatives that the systems idea regains much of its credibiiity. For while there have been a significant number of critical reactions to the difficulties encountered in "rationally"[1] attempting to solve social problems, few critics have posed worthwhile alternatives. Some have simply thrown up their hands while awaiting a radical transformation of society and/or its value system. Others have adopted rather naive approaches, perhaps in ignorance of the alegedly rational techniques of problem-solving. Braybrooke and Lindblom identify a "naive criteria method" in which the assertion of a few general values or goals is assumed to provide sufficient mechanisms of choice. Implicit here is the use of little or no analysis—rational or otherwise.[2] This is similar to what Hitch and McKean have called a "requirements approach" in which problem-solving is based on need alone without a comparison of alternative means, and especially costs, of achieving the defined needs.[3] Both groups also identify a "priorities approach"[4] in which the decision-maker simply ranks alleged needs according to their perceived relative urgency.

As an example of this type of thinking in urban planning, we have the work of William Michelson.[5] He asserts that in their design processes, planners and designers have not really accounted for people's "needs," which he identifies in terms of life-style and position in a life cycle; he is oblivious to questions of public resource allocation, market and political forces and, in essence, problem-solving. His assertion of perceived need does not generate any methodological tools for its satisfaction.[6]

In architecture we have various naiveties also, as for example, Venturi's Las Vegas technique [7] that asserts the correctness of what is "almost alright," in an ambience of complexity and contradiction. There are also the architectural manifestos of the so-called "Five." While adherents of both theories have on occasion produced worthwhile architecture, one legitimately wonders if there is any relationship between this fact and their theorizing—the essential theoretical content being much too trivial to account for the enormous complexity of problem-solving for today's architecture and urban planning.

**Muddling Through with Disjointed Incrementalism.** When one reviews the literature there seems to be only one seriously developed alternative: Charles E. Lindblom's concept of disjointed incrementalism. It alleges to describe, moreover, what people actually do when confronted with complex problem-solving.[8]

Lindblom's thesis was first presented in 1959 as a view of the way one really "muddles through" problem-solving. He contrasted his approach (also then known as one of Successive Limited Comparisons or the "Branch" Method) with the Rational-Comprehensive (or "Root" Method) in the following terms:

| Rational-Comprehensive (Root) | Successive Limited Comparisons (Branch) |
|---|---|
| 1a Clarification of values or objectives distinct from and usually prerequisite to empirical analysis of alternative policies | 1b Selection of value goals and empirical analysis of the needed action are not distinct from one another but are closely intertwined |
| 2a Policy-formulation is therefore approached through means-end analysis: First the ends are isolated then the means to achieve them are sought | 2b Since means and ends are not distinct, means-end analysis is often inappropriate or limited |
| 3a The test of a "good" policy is that it can be shown to be the most appropriate means to desired ends | 3b The test of a "good" policy is typically that various analysts find themselves directly agreeing on a policy (without their agreeing that it is the most appropriate means to an agreed objective) |

| Rational-Comprehensive (Root) | Successive Limited Comparisons (Branch) |
|---|---|
| 4a Analysis is comprehensive: every important relevant factor is taken into account | 4b Analysis is drastically limited:<br>   i) important possible outcomes are neglected;<br>   ii) important alternative potential policies are neglected;<br>   iii) important affected values are neglected |
| 5a Theory is often heavily relied upon | 5b A succession of comparisons greatly reduces or eliminates reliance on theory [9] |

Later with David Braybrooke, Lindblom developed a fuller account of the method—now termed Disjointed Incrementalism. In it, one proceeds in problem-solving on the basis of margin-dependent choice with a focus "on the increments by which the social states that might result from alternative policies differ from the status quo." [10] Both the number and the variety of policy alternatives considered by a planner are greatly restricted and the evaluation of probable consequences of each relevant alternative is also severely limited.[11] Thus, the attempt is to portray only a few social states ("states of the system") which differ only slightly from each other and from the existing system state:

Incremental choice . . . [of] policies proceeds through comparative analysis of no more than the marginal or incremental differences in the consequent social states rather than through an attempt at more comprehensive analysis of the social states . . . [and] choice among policies is made by ranking in order of preference the increments by which social states differ.[12]

Up to this point, the method may seem to be simply a limiting cast of the synoptic or systems ideal. A more fundamental attack of the logic of the latter, however, is apparent in the view that ends can govern means since:

(1) The analyst chooses as relevant objectives only those worth considering in view of the means actually at hand or likely to become available; (2) He automatically incorporates consideration of the costliness of achieving the objective into his marginal comparison, for an examination of incremental differences in value consequences of various means tells him at what price in terms of one value he is obtaining an increment of another; (3) While he contemplates means, he continues at the same time to contemplate objectives, unlike the synoptic analyst who ideally must at some point finally stabilize his objectives and then select the proper means.[13]

In the continuing evolution of this position he asserts that the "intelligence of a democracy" lies in the fact that within it decisions are largely based upon consensus. Implicit is the tautology that the best decisions are necessarily consensual. To achieve this consensus, partisans to a decision interact (viz: bargain, contend, manipulate, confront, negotiate, and so forth) in pursuit of their self-interest. This interaction results in decisions which reflect the relative power and persuasiveness of the contenders in a process of mutual adjustment among the contending partisans.

In many respects, Lindblom's model is attractive. Political choice ameliorates the fact-value problem since political democracy implies participation by the citizens (and, by extension, their value systems) in decision-making, either directly or through their surrogate representatives. Via such a mechanism, the aggregate of public values on an issue can somehow be expressed and at least crudely assessed.

As Etzioni has pointed out, perhaps overemphatically, the synoptic or systems ideal is not really suitable for problems more complicated than the planning of traffic lights because of the demands it places upon the decision-makers' limited cognitive capacities. Lindblom himself has added cogent specifics to these limitations. His own model is certainly open to question. The specifics of criticism can focus either upon its "disjointed" or on its "incremental" character. On the former issue, Lindblom himself provides the prologue for the indictment:

> Analysis and evaluation are disjointed in the sense that various aspects of public policy and even various aspects of any one problem or problem area are analyzed at various points, with no apparent coordination and without the articulation of parts that ideally characterizes subdivision of topic in synoptic problem solving. Of course, analysis and evaluation are in a secondary sense also disjointed because they focus as heavily as they do on remedial policies that "happen" to be at hand rather than addressing themselves to a more comprehensive set of goals and alternative policies.[14]

We have unfortunately experienced the results of remedial policies that "happened" to be on hand in addressing housing, transportation, pollution, and other urban problems. Fragmented and disjointed policy resulting in uncoordinated public and private action has, as often as not, resulted in exacerbation of the problems. In short, many urban systems are in fact holistic and require a close coordination of policies and programs if they are to function efficiently or somehow in "the public interest." For example, clean air is rapidly becoming a "public good" as opposed to a "ubiquitous good" in the economic sense; it is in a consumer's interest to understate his valuation of this type of good. Also, a few polluters or consumers of air acting independently (i.e., in a fragmented or disjointed fashion) to improve its quality will have virtually no effect, allowing little hope for the prospect for "mutual adjustment" of the issue by disjointed "partisans." Such a problem *demands* central coordination to assert the collective interest and to apportion costs both to the polluters and to the public as a collectivity; "partisans" will not do this.

Moreover, the central coordinator or planner must plan rather synoptically if even he is to have an effect. With respect to housing, transportation, and other urban systems the impact of proliferating agencies and programs as they attempt to meet the problems unsynoptically—by focusing on their own specific administrative turf—is disconcerting. The problems remain, and the public receives added costs due to burgeoning bureaucracies and a lack of coordination which often results in the establishment of different and potentially contradictory policies.

In sum, the Lindblom thesis is largely blind to the systemic nature of society and the city; hence he asserts that the problems of these systems can be dealt with in a disaggregated, piecemeal fashion. While there are certain dangers in overstating the holistic orientation, yet to go to the opposite position and see the city as a disjointed congeries of noninterdependent parts is an even greater misconception. The suburb versus central city "partisan mutual adjustment" through political contention is only too vivid. Reason—synoptic or otherwise—informs us that interdependencies *do,* in fact, exist. Yet, precisely because of the partisan nature of the conflict, these same interdependencies are ignored with the resulting deferment of solutions to metropolitan problems. Meanwhile, the problems get worse.

Limitations on rationality are also imposed by the incremental nature of Lindblom's model. Here one can only wonder how incremental urban decisions can be. For an example of this, we might look at the planning of a new line for an existing subway system—by definition an incremental decision. This increment will generate many interactions with other urban systems. For one thing, when the line goes in, it will affect land values and the ultimate character of land use along its path; the transformation in land use will itself have further repercussions. Thus, an act which initially differed only incrementally from the status quo will have numerous long range effect on the urban system; the consequences of which, both intended and unintended, should have been anticipated, however sketchily, at the outset of the incremental change. Is it utopian to assert that had this problem been addressed synoptically or holistically perhaps the interdependences of the problematic elements would have been perceived, the potential consequences examined and, just possibly, a better solution achieved?

Another aspect of the model's incremental quality is its conservative bias.[15] When you propose to change incrementally from the status quo, you are by definition conservative. Given the nature and extent of our urban problems, it does not seem terribly radical to suggest that this bias might suppress more wide ranging and innovative thinking about them.

Some of these limitations of Lindblom's model as a guide to problem-solving or decision-making are apparent in his own definition of its relevance. He has related various decision-making approaches to the planner's understanding of the problem and to the degree of change involved. On this basis, his model is seen as useful only in situations involving low understanding and small change.

The final and perhaps most fundamental issue raised by the model is the prospect of means determining ends. One is reminded of the oft-quoted epigram of Einstein's that the central malaise of our age lies in the fact that we are so much concerned with means that we give little thought to the ends to which these means are to be put. In sum, then, Lindblom's model is not so much an approach to planning as it is an attack upon it; its plea for essentially political choice seems to exclude or preclude the possibility of meaningful planning.

**Middle Positions.** Not surprisingly we find some well-articulated positions between the excessive rationality of the systems ideal and the nonplanning of disjointed incrementalism. The most interesting of these middle positions are Simon's "satisficing" and Etzioni's "mixed scanning" approaches.[16] The degree to which each position approaches the synoptic determines the advantages and disadvantages of that approach; as it does similarly with respect to disjointed incrementalism.

Simon's satisficing model developed in response to the conception of economic man as a useful construct for the organizational decision-maker. To bring economic man into the real world, Simon recognizes a "principle of bounded rationality." This amounts essentially to a compilation of many of the elements encountered in our critique of the systems ideal: the inability of the decision-maker to acquire all the facts on the issue, his limited resources (time, money, information, intellect, etc.) which inhibit a synoptic view, and in his failure to distinguish between fact and value in a problem situation.

Because of the decision-maker's limited capacity for rationality, Simon says he simply cannot optimize in the context any complex decision, no matter how hard he tries. Instead, he must "satisfice," that is, look for a solution which is good enough as opposed to optimal. To use Simon's classic expression for the model: it means one does not look for the sharpest needle in a haystack, only for a needle sharp enough to sew with. Thus, a satisficing planner "has no need of estimates of joint probability distributions, or of complete and consistent preference orderings of all possible alternatives of action."[17] In fact, he does not even examine all the options or substitution possibilities, only those that look promising: intuition and judgment are allowed back into a complex world.

Etzioni's approach is similar to Simon's. Again, it is proposed because of the unrealistic demands of the synoptic or systems ideal and the potential irrationalities of a disjointed incrementalist approach. Etzioni's definition of mixed scanning is somewhat obscure:

Actors whose decision-making is based on a mixed-scanning strategy differentiate contextuating (or fundamental) decisions from bit (or item) decisions. Contextuating decisions are made through an exploration of the main alternatives seen by the actor in view of his conception of his goals, but—unlike what comprehensive rationality would indicate—details and specifications are omitted so that overviews are feasible. Bit-decisions are made "incrementally" but within the contexts set by fundamental decisions (and reviews). Thus, each of the two elements in the mixed-scanning strategy helps to neutralize the peculiar shortcoming of the other: Bit-incrementalism overcomes the unrealistic aspects of comprehensive rationalism (by limiting it to contextuating decisions), and contextuating rationalism helps to right the conservative bias of incrementalism. Together they make for a third approach which is more realistic and more transforming than each of its elements.[18]

As I see it, this model varies from Simon's satisficing approach only in that it prescribes a way of satisficing: by scanning. This means that the decision-maker looks over the entire matrix of elements in the problem situation quite generally or superficially at first. He gleans from this view a measure of the heightened significance of some elements as opposed to the others. This scanning enables him to pick out the significant elements of the problem situation and thus to focus in greater detail upon them. He does not view all options as the synoptic idealist would. He only rationally—systematically—examines those options and elements which previous scanning has indicated as significant. Etzioni uses the analogy of a weather satellite with various cameras; it has a broad-view camera with which to see the general overall pattern and it has another narrower-view camera which seeks out areas of specific meteorological interest so as to provide much sharper focus for greater detail in the area of concern.

## 12. Summary and Synthesis

Given some sense of the probable utility of the systems ideal as opposed to "disjointed incrementalism" and of the apparent reasonableness of "satisficing" and "scanning" relative to these polar models, it would be easy to conclude that virtue somehow lies in the middle. It also must be admitted that to a certain extent this is recognized in practice.

Something must be said, however, about the scale and scope of decisions and about the contexts within which they are taken. A basic theoretical distinction can be made between long-range, large-scale planning and specific, detailed programs on a smaller scale and for the near future which have their genesis in broader contextual planning.[1] This suggests a planning process that is sensitive both to Etzioni's contextuating distinction and Simon's "factorization" of planning in terms of a hierarchy of decisions.[2]

This distinction between overall planning and specific or detailed plans has significant implications for the structuring of thought in urban planning. The traditional architectural design process is useful here as an illustration. In it, at the initiation of a program during design schematics, the process is quite fluid and the architect is usually open to suggestion on all points of the design, even encouraging a fuller exploration and articulation of objectives by the client. The architect attempts to draw out of the client the latter's perception of "needs" for the building—the performance characteristics he seeks and what he envisions as the end product. This often results in a redefinition of the means (i.e., money) that are available to the client in achieving the objectives.

To this point, the process might suggest disjointed incrementalism as the appropriate mode of procedure; a fluid decision-process which is more or less democratic with the encouragement of an open discussion of goals and of means. Radical change can be accommodated readily by the architect at this early stage. When the architect moves on to the preliminary working drawings, he is somewhat less amenable to substantial change as he has begun to solidify his concept of the major system elements and their relationships. As he continues further on in the working drawings, he becomes even less tolerant of change simply because his problem-solving or decision process now demands it. He is now working synoptically—and "rationally"—the context has now been established and his system is more tightly structured. The complexities generated by the interdependence of the system elements do not allow for a general restructuring of the overall system except through the costly reiteration of the design process already accomplished; theoretically this may be desirable but in practice it is almost inevitably avoided.

Finally, when the specifications and working drawings are drawn up and the actual construction begins, the architect adamantly resists change. Thus we can see in the development of a project that at the outset there is much fluidity in the thinking process and that this fluidity congeals gradually over time, ultimately becoming rigidly fixed as the end product of the thought process is approached. There is also a relationship between the level of abstraction of the thought and the specificity of the design being produced. The techniques used to support the thinking process also change from loosely structured conversations at the outset to precise mathematical modeling of structural and other systems toward the end of the process.

An analogous planning mode has been suggested by a planning advisory group to the British government.[3] The study was not generated, however, by the theoretical considerations which have loomed large here: to wit, the nature of the planning process, its conflict and overlap with market and political process, and so forth. It arose from the practical necessities of planning in England and was informed by everyday experience with the planning process. The advisory group suggested a fitting of the nature of the planning effort to the scale and scope of a decision-maker's responsibilities. General policy matters, decided at the ministerial level, are incorporated in a broad policy plan which lacks implementing detail. Then, as one moves down in the hierarchy of decision responsibility (i.e., local governments, housing authorities, etc.), the plan acquires greater detail and could employ more rigorous (i.e., systems analytic) processes in its definition.

Such a conception of planning meets several of the objections to the synoptic or systems ideal: limitations imposed by a decision-maker's potential for rationality, including his ability to acquire facts, determine objectives, establish trade-offs among objectives, analyze the available options, determine the consequences of each alternative and, ultimately, delineate the preferred alternatives. At the city-scale, this task would be impossible except at a "satisficing" level. Even assuming an improbable omniscience on the part of the planner to the extent that he *could* go through such an analysis, the time constraint alone would not permit it. By the time the analysis was accomplished, a new set of problems, objectives, and political constraints would exist.

This is not abstract conjecture but a realistic depiction of the problem-solving ambience within which the urban planner must work—an account of certain realities one tends to forget when engrossed in the analytic process. An attempt to sum up these realities is made in Figure 55 delineating a potentially useful structure for the city-planning process. It reflects the fact that large social systems are very much in flux; they are dynamic and stochastic as opposed to being neatly deterministic. While probability theory can no doubt make manageable some of the phenomena exhibited by these systems, it is absurd to think that there exists, or perhaps could exist, a model which would neatly tie together all the disparate phenomena with which the urban planner and architect must deal. The conditions within their physical systems change over time as do the values of the various actors concerned with improving the systems; as a result, definitions of problem situations will change as social, economic, and political influences on the systems change. These influences are generally outside the planner's span of control and perhaps even of his vision. Moreover, there is an ultimate relativism in the choice of objectives one seeks in bringing about change in the built environment; regardless of one's neo-Enlightenment tendencies—faith in a society's ability "rationally" to solve social problems—one must recognize that the ultimate decisions on ends sought and the trade-offs established between them are formulated within the political arena. This formulation is often unclear. As one informed and experienced urban analyst has put it:

Problem areas, like housing, and welfare, and even police work decline to provide analysts even the entrance point of an accepted set of

90

objectives. In some cases, objectives are disputed; in others, they are ignored. In either event, the analyst lacks any authoritative statement of what it is that must be maximized, or minimized, and under what constraints. . . . Urban governments persist in avoiding any close identification of objectives. They must. The purposes of some programs are cynical; the objectives of many more are multiple, and maintaining disparate sources of support for them requires ambiguity. Moreover, men who run for office know far more poignantly than researchers do the odds against getting change actually accomplished. They know, therefore, that large promises made with specificity are invitations to proof of failure two or four years later. Yet large promises must be made. Hence vagueness.[4]

To persist in the zealous pursuit of an ideal of technique and process—systems or otherwise—in the light of these conditions suggests naivete or cynicism. Perhaps, as Yeats once said, it is merely an attempt of the will to do the work of the imagination. In any event, it often has the quality of a ritual act, something one does when confronted with unsolvable problems. D. N. Michael calls it "ritualized rationality" which

consists of undertaking elaborate rational logical or mathematical exercises which are often not used or useful. They are often not used because they weren't intended to be used in the first place—their purpose was a magical, go-to-church type, comforting ritual for the decision makers and their model makers, who, confused and frustrated by the state of the world, turn for the answers to "logic" as a faith. These rational exercises are often not useful, because the models, to be rational or mathematical, usually require a denaturing of reality to the point where the conclusions reached are not reasonably applicable to the complexities of the world.[5]

This ritualistic precision is partly the result of a trade-off between the systems idea's central concerns of holism and rationality. The precise and cogent analysis that one associates with the better systems studies is often accomplished at a cost of excluding values other than the economic or quantifiable from a problem's definition. Here the irony is that the systems approach becomes insufficiently holistic in order to achieve the high degree of rationality associated with it; one must narrow, atomize, and ultimately discard many of the values which could inform the analytic process so as to produce a technically correct or "exacting" analysis of the problem. The result is often a manipulation of technique via methods both mechanical and illusory with a focus on the trivial and external and applied with an uneasiness or disregard for the substance and meaning of an issue; in sum, a denaturing of reality.

It was their search for a more complex and coherent definition of reality that led the early organic architects to pursue their analogy. The old idea of finding system in nature and emulating its suggestive forms had currency among a significant number of architects and undeniably contributed to the development of modern architecture. Perhaps, as we have noted, it can even be viewed as a naive antecedent of what we now know as bionics. Moreover, this vision served a multiplicity of values, some of which were aesthetic. For example, one of the incentives to idealize the small cellular community was the aesthetic sense of completeness it generated; the design had coherent form, it was immediately imageable in the mind and, most significantly, it was a complete aesthetic statement. It was, in sum, a complete and coherent work of art. Never mind the fact that it may have been economically or perhaps socially dysfunctional, there was an aesthetic satisfaction; aesthetic values were served even if economic ones were neglected.

While one might question the relative significance of the values served, the important matter with respect to the architects' problem in giving form to things was that the early and essentially architectural use of the organic analogy with its holistic manner of looking to natural systems for models of form was enormously suggestive. It provided the architect with a wide array of potential organic or biomorphic forms—a vocabulary of form—expressive of holistic entities: coherent wholes. These forms—when used sensitively—could furnish models for the aggregation of architectural parts into similarly holistic entities. However, use of the organic idea by architects did not really go beyond the merely suggestive; especially so in the writings of the principal users, Sullivan and Wright. A suggestion was deemed sufficient; the desire to elaborate or to develop specific techniques in the manner of systems analysts was simply not there.

It is here that modern systems theory comes into its own: it has produced useful and necessary techniques for problem-solving. An architect would no longer attempt to intuit the shearing stress or bending moment in a beam since exact mathematical formulations for determining these stresses exist. While some quibble about presumed isomorphism and similar problems might exist in specific cases, the mathematically derived measure of stress closely approximates that obtained in real world tests of the beam. It is not unreasonable to presume that similar formulations can be made to analyze conditions in systems larger than a building's structure; perhaps in entities as large and as complex as cities or societies. In fact, many such mathematical techniques have been developed in economics, operations research, and various systems areas.

This development of technique is seductive, however. Because of its relative sophistication vis-à-vis traditional and intuitive architectural formulations, one is inclined, almost unconsciously, to premise the entire decision-making process on it. Hence the significance of economic technique and rationality in social problem-solving: it works, it gives quantified answers, and other trades have yet to develop similar operational methodologies. Other values, being largely intuitive or at least unquantifiable, are much more nebulous, and hence difficult to deal with in the problem-solving process; they are not as easily incorporated in the ultimate plans and designs. The seduction of economic method is hazardous, though. Almost by definition, the techniques involved demand a high degree of quantification. Among other things, this may leave out the "quality" or the aesthetic attributes of things—a principal concern of the architect. While seduced by technique then, one must realize that the ultimate search is for a technique with which to manage technique.

In this context, the systems approach is something of an ideal; it is an approach that has proved enormously useful in the planning and design of "hard" systems within which conditions can be quantified and objectives specified along with the necessary trade-off of values that will define the optimal. The previous assessment provided ample recognition of the fact that the realities of planning and architecture are not necessarily amenable to definition and analysis through the simple adaptation of techniques used in these more rigorous systems areas. Yet going to the polar opposite of disjointed incrementalism, or "learning from Las Vegas," merely brings about more of the extant—which is not necessarily good. The systems approach is holistic and rational and, despite the difficulties, when used with sensitivity and appropriate caution concerning its limitations, it can work. In sum, it is an enormously suggestive idea; an idea abused by many: in its early organic days, with excessive biological baggage; today, in its systems format, by the "terrible simplifiers" who see in it easy mechanisms for the description and analysis of very complex social processes which will be dealt with, in turn, often by simplistic techniques. Notwithstanding this, the historical recurrence of the systems idea testifies to its continuing significance. As Clovis Heimsath puts it:

Many trends beckon in this time of rapid change within the profession, a period that might be termed "postmodern" architecture. There are inflatable structures, Las Vegas zip-zap instant design, convoluted geometry and the systems process.

Is the systems process misunderstood and underestimated? Perhaps . . . [yet] the systems approach will outlive today's other trends and will be the herald of tomorrow's architectural practice. . . . A rush to variety and non-order can only be meaningful against the tension of an ordering.[6]

In summing up, we might benefit from an insight provided by Henri Lefebvre, in introducing Philippe Boudon's assessment of Le Corbusier's Quaritiers Modernes Fruges at Pessac. He says that the architect and planner perceive

different levels of reality and different levels of thought. . . .

(a) First there is the *theoretical level,* at which theory tends to merge with ideology or, to be more precise, is not usually sufficiently distinquished from ideology. . . . Architects and town planners . . . deal with empirical problems by reference to town-planning ideologies. . . .

(b) Then there is the *practical level,* at which ideological considerations are supplemented by other, quite different factors. Here the architect exercises his mind and his will, bringing them to bear on the practical needs of the future occupants. Some of these needs are clearly recognized, others are not. And so Le Corbusier's architectural practice is seen to be more hesitant, more flexible and more vital than his architectural theory. But both ideological and theoretical considerations are forced to give way in the face of reality.

(c) Finally, there is the *town-planning level,* at which a certain way of life, a certain style (or absence of style) makes itself felt. The social activities of individual occupants and groups of occupants, which have been influenced to a greater or lesser degree by the different groupings within the district, are seen for what they are. At this level we find a specific topology, a concrete rationality that is more impressive and more complex than abstract rationality.[7]

Although it may seem to some a contradiction in terms, I think it was a search for this concrete and more impressive rationality that stimulated the organic school of architects in their exploration of the analogy. This exploration, however, was largely at Lefebvre's "theoretical level" and usually within the ideological context noted here. More significantly, it did not result in a useful theory. The writings of Sullivan and Wright, for example, are essentially a series of insights—perhaps profound insights—into the planning and design process, but it is difficult to argue that there is a coherence or integration in these works that would in any way constitute an adequate theoretical foundation. Moreover, there was often excess and abuse in the use of the organic analogy.[8] This theoretical insufficiency is reflected in the inability of the followers of Sullivan and Wright to build—in any substantial way—on their contribution.

Continuing in Lefebvre's taxonomy, systems analysis is eminently at the "practical" level; perhaps too much so. However, while the systems analytic process may have significant shortcomings when viewed in the context of actual practice today, there are no attractive alternatives: the systems analyst and his operations research colleagues have a kit of tools that work; tools that the aesthetically oriented problem-solver needs. Whether or not the latter has a perception of that need is another matter. Perhaps a diminution of the analyst's fondness for the economically "rational" would allow him to appreciate the values which stimulate the architect and others. The result could be a much more creative planning and design process: one which, among other things, incorporates the suggestive form-giving insights architects have perceived in natural systems that act holistically. This would bring the idea full circle back to a concern with organic architecture and natural systems as the inevitable examplar in the mode of Sullivan, Wright, the Metabolists, et al. Perhaps this is simply to suggest that the systems subculture of bionics provides an attractive theoretical avenue for the architect and planner concerned with the form of things; minimally, it's an avenue worth exploring.

To some, this might seem a regression; to others an enormous expansion of the potential inherent in the systems concept; an expansion that could not be realized in the earlier and more limited use of the idea by Wright and the organic school. This would move the systems approach beyond its present, nearly exclusive, concern with the analytic function into a more synthetic or creative area: to the central problem of generating the options or alternative systems—giving form to the built environment.[9] While this is generally thought of as an extension of the analytic process, in fact, very little thought is given to how it is actually done. Having an option, many techniques exist to analyze it in great detail—with the reservations previously noted—but generating that option is something largely alien to the approach itself. Louis Kahn went to the heart of the problem:

> The creation of art is not the fulfillment of a need but the creation of a need. . . . The world never needed Beethoven's Fifth Symphony until he created it. Now we could not live without it.[10]

This generation of a new need that was not previously anticipated or sought is central to the work of the architect. As Louis Sullivan said, it *is* his art, and this is what giving form to the built environment is ultimately all about.[11] One may analyze, one may assess, one may determine the cost-reward structure of various options, but ultimately one must create. As an aid in this creative process, a congeries of techniques will not provide either a sufficient intellectual framework or the coherence essential to the process. Nor will the limited perspective of economic rationality supply an adequate value-orientation. However, the process does need the "partisan efficiency advocate" that Schultze suggests,[12] and it is largely via economics, operations research, and systems engineering that we currently have analytic techniques to determine "efficient" solutions in architecture and planning. But, while we must have the partisan efficiency advocate, this advocate cannot be allowed to determine both context and procedure of the problem-solving process; procedures which, by default, largely determine the ultimate outcomes. However, with respect to the architectural theory, the use of metaphor and analogy—while suggestive—was also ultimately insufficient in the problem-solving process.

In this context it seems reasonable to suggest that the efficiency advocate and the architectural theorist might build upon their common foundation in a holistic approach: the former with his kit of tools, the latter with his sensitivity to the "art" of building the physical environment. A coalition could be fruitful; although it may not result in Beethoven's Fifth it should give us something better than current work.

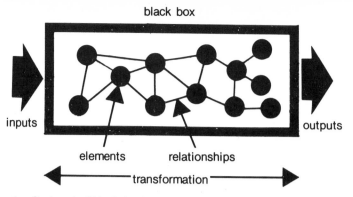

black box

inputs

outputs

elements

relationships

transformation

1.  System in "black box" format. *Illustrations for which no source is given are the author's.*

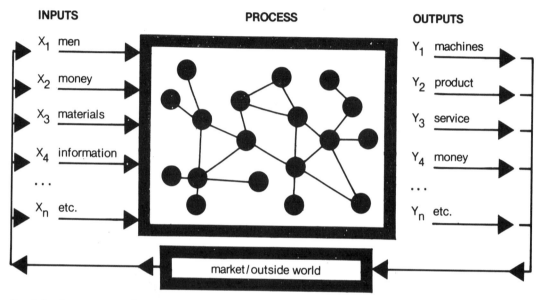

**INPUTS**

$X_1$ men

$X_2$ money

$X_3$ materials

$X_4$ information

...

$X_n$ etc.

**PROCESS**

**OUTPUTS**

$Y_1$ machines

$Y_2$ product

$Y_3$ service

$Y_4$ money

...

$Y_n$ etc.

market/outside world

2.  A business as a system.

3.  Dahl's representation of Parson's Social System. Robert A. Dahl, *Modern Political Analysis,* Englewood Cliffs: Prentice Hall, 1965.

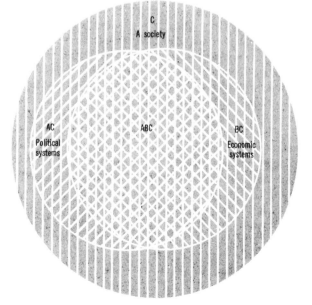

C
A society

AC
Political
systems

ABC

BC
Economic
systems

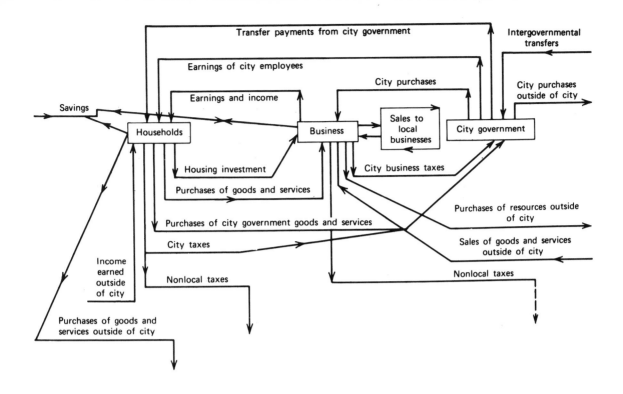

4. The city system as reflected in its money flows. William L. Henderson and Larry C. Ledebur, *Urban Economics: Problems and Processes,* New York: Wiley, 1972.

5. Abstraction and model-building.

6. Systems analysis and the planning process. Bernard H. Rudwick, *Systems Analysis for Effective Planning,* New York: Wiley, 1969.

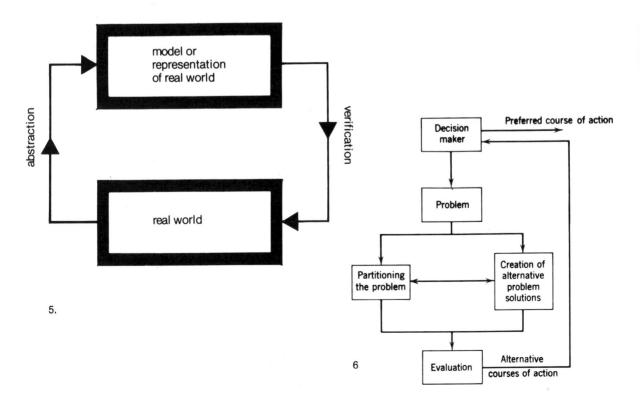

5.

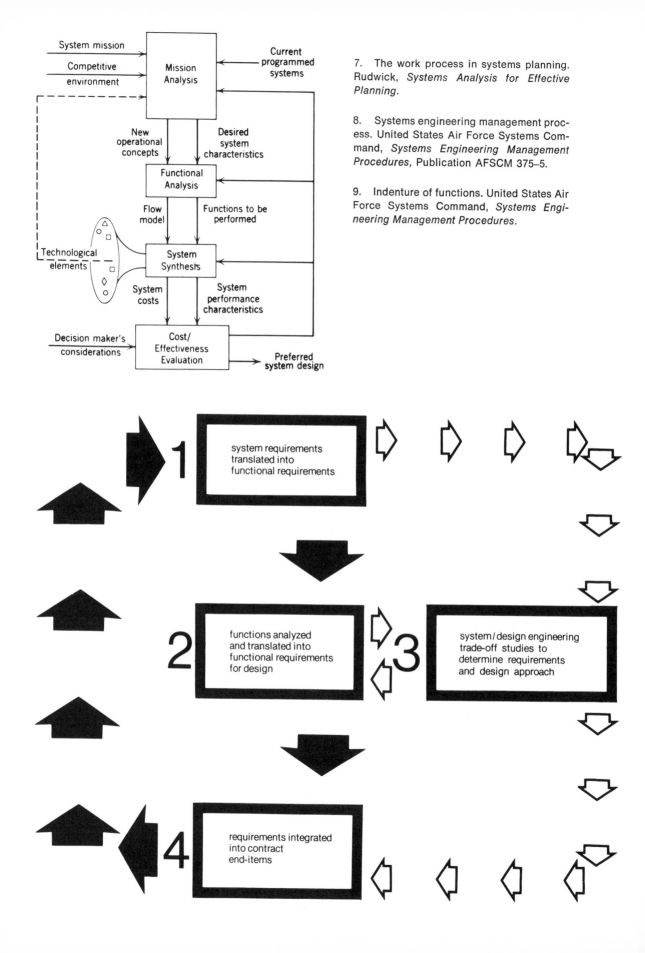

7. The work process in systems planning. Rudwick, *Systems Analysis for Effective Planning.*

8. Systems engineering management process. United States Air Force Systems Command, *Systems Engineering Management Procedures,* Publication AFSCM 375–5.

9. Indenture of functions. United States Air Force Systems Command, *Systems Engineering Management Procedures.*

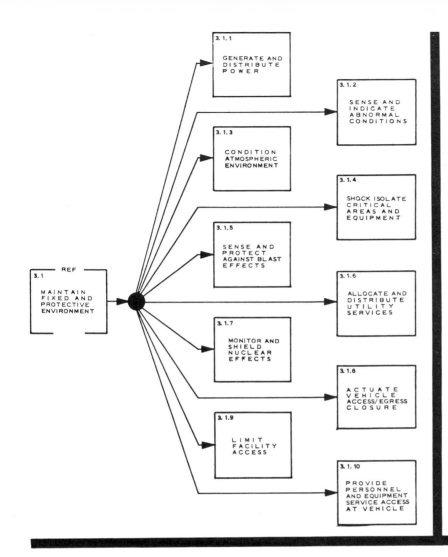

In the diagram on the left, several functions of a hypothetical system of facilities to support an aero-space system are illustrated. These functional definitions are then further refined in a continuing analysis of more detailed functions within the larger or gross functional definition. For example, the diagram below continues the functional definition described briefly in 3.1.1. on the left. The illustrations provided are 2nd and 3rd level functional specifications in a methodology which requires a finite but extensive chain of continuing levels of functional definition leading ultimately to a determination of the form of physical systems.

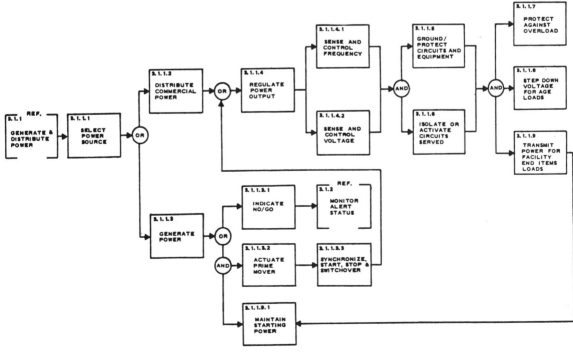

10.

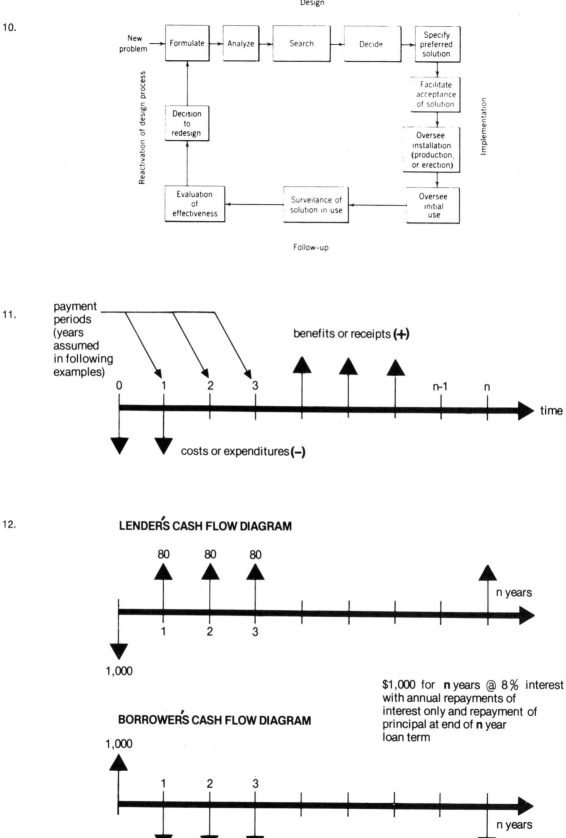

Design

New problem → Formulate → Analyze → Search → Decide → Specify preferred solution

Reactivation of design process

Decision to redesign

Facilitate acceptance of solution

Implementation

Oversee installation (production, or erection)

Evaluation of effectiveness ← Surveillance of solution in use ← Oversee initial use

Follow-up

11.

payment periods (years assumed in following examples)

benefits or receipts (+)

0    1    2    3              n-1    n    time

costs or expenditures (−)

12.

**LENDER'S CASH FLOW DIAGRAM**

80    80    80

n years

1    2    3

1,000

$1,000 for **n** years @ 8% interest with annual repayments of interest only and repayment of principal at end of **n** year loan term

**BORROWER'S CASH FLOW DIAGRAM**

1,000

1    2    3

n years

80              1,080

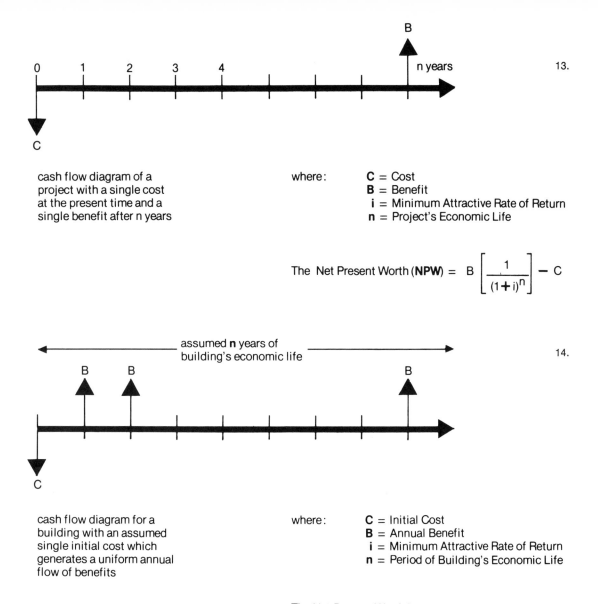

cash flow diagram of a
project with a single cost
at the present time and a
single benefit after n years

where:  **C** = Cost
**B** = Benefit
**i** = Minimum Attractive Rate of Return
**n** = Project's Economic Life

$$\text{The Net Present Worth (\textbf{NPW})} = B \left[ \frac{1}{(1+i)^n} \right] - C$$

cash flow diagram for a
building with an assumed
single initial cost which
generates a uniform annual
flow of benefits

where:  **C** = Initial Cost
**B** = Annual Benefit
**i** = Minimum Attractive Rate of Return
**n** = Period of Building's Economic Life

The Net Present Worth is

$$NPW = B \left[ \frac{(1+i)^n - 1}{i(1+i)^n} \right] - C$$

10. The design cycle. Edward V. Krick, *An Introduction to Engineering and Engineering Design*, New York: Wiley, 1965.

11. The cash-flow diagram.

12. Borrower's and lender's cash-flow diagrams for same loan.

13. Present worth of a single cost resulting in a single benefit.

14. Present worth of a single cost with a uniform stream of annual benefits.

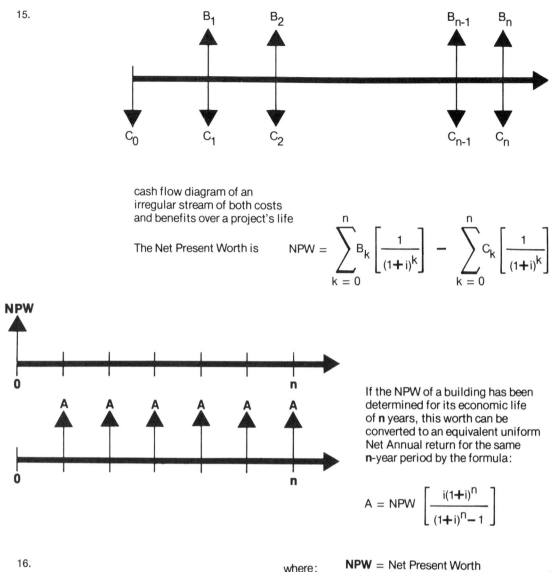

15.

cash flow diagram of an
irregular stream of both costs
and benefits over a project's life

The Net Present Worth is

$$NPW = \sum_{k=0}^{n} B_k \left[ \frac{1}{(1+i)^k} \right] - \sum_{k=0}^{n} C_k \left[ \frac{1}{(1+i)^k} \right]$$

**NPW**

0                n

If the NPW of a building has been
determined for its economic life
of **n** years, this worth can be
converted to an equivalent uniform
Net Annual return for the same
**n**-year period by the formula:

$$A = NPW \left[ \frac{i(1+i)^n}{(1+i)^n - 1} \right]$$

16.

where:
  **NPW** = Net Present Worth
  **A** = Equivalent Uniform Net Annual Return
  **i** = Minimum Attractive Rate of Return
  **n** = Period of Building's Economic Life

15.   Present worth of an irregular stream of costs and benefits.

16.   Conversion of a net present worth to its equivalent net annual return.

17.   Simple benefit-cost ratio.

18.   Lichfield's planning balance sheet. Morris Hill, "A Goals-Achievement Matrix
for Evaluating Alternative Plans," in *Journal of the American Institute of Planners*,
January 1968.

19.   Hill's goals-achievement matrix. Hill, "A Goals-Achievement Matrix."

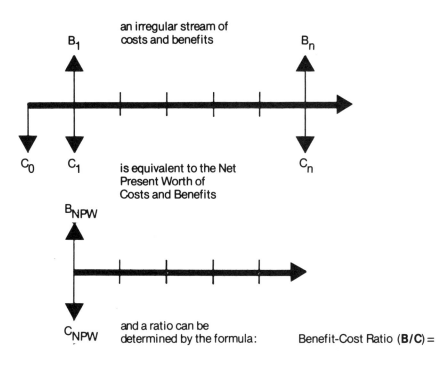

## 17

B₁ — an irregular stream of costs and benefits — Bₙ

$B_1$ ... $B_n$

$C_0$ ... $C_1$ — is equivalent to the Net Present Worth of Costs and Benefits — $C_n$

$B_{NPW}$

$C_{NPW}$ — and a ratio can be determined by the formula:

Benefit-Cost Ratio (**B/C**) =

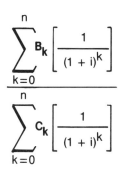

$$\text{Benefit-Cost Ratio }(\mathbf{B/C}) = \frac{\displaystyle\sum_{k=0}^{n} B_k \left[\frac{1}{(1+i)^k}\right]}{\displaystyle\sum_{k=0}^{n} C_k \left[\frac{1}{(1+i)^k}\right]}$$

|  | Producers | | | | | | | |
|---|---|---|---|---|---|---|---|---|
|  | Plan A | | | | Plan B | | | |
|  | Benefits | | Costs | | Benefits | | Costs | |
|  | Cap. | Ann. | Cap. | Ann. | Cap. | Ann. | Cap. | Ann. |
| X | \$a | \$b | — | \$d | — | — | \$b | \$c |
| Y | $i_1$ | $i_2$ | — | — | $i_3$ | $i_4$ | — | — |
| Z | $M_1$ | — | $M_2$ | — | $M_3$ | — | $M_4$ | — |
|  | Consumers | | | | | | | |
| X¹ | — | \$e | — | \$f | — | \$g | — | \$h |
| Y¹ | $i_5$ | $i_6$ | — | — | $i_7$ | $i_8$ | — | — |
| Z¹ | $M_1$ | — | $M_3$ | — | $M_2$ | — | $M_4$ | — |

| Goal description: | $\alpha$ | | | $\beta$ | | | $\gamma$ | | | $\delta$ | | |
| Relative weight: | 2 | | | 3 | | | 5 | | | 4 | | |
| Incidence | Relative weight | Costs | Ben. | Relative weight | Costs | Ben. | Relative weight | Costs | Ben. | Relative weight | Costs | Ben. |
|---|---|---|---|---|---|---|---|---|---|---|---|---|
| Group a | 1 | A | D | 5 | E | — | 1 | | N | 1 | Q | R |
| Group b | 3 | H | | 4 | — | R | 2 | | — | 2 | S | T |
| Group c | 1 | L | J | 3 | — | S | 3 | M | — | 1 | V | W |
| Group d | 2 | — | | 2 | | — | 4 | | | 2 | — | — |
| Group e | 1 | — | K | 1 | T | U | 5 | | P | 1 | — | — |
|  | | Σ | Σ | | | | | Σ | Σ | | | |

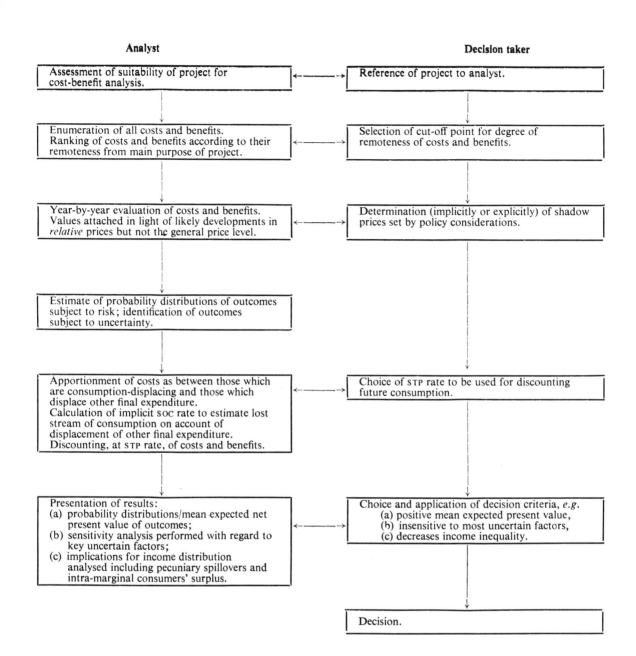

**Analyst**

| |
|---|
| Assessment of suitability of project for cost-benefit analysis. |

| |
|---|
| Enumeration of all costs and benefits. Ranking of costs and benefits according to their remoteness from main purpose of project. |

| |
|---|
| Year-by-year evaluation of costs and benefits. Values attached in light of likely developments in *relative* prices but not the general price level. |

| |
|---|
| Estimate of probability distributions of outcomes subject to risk; identification of outcomes subject to uncertainty. |

| |
|---|
| Apportionment of costs as between those which are consumption-displacing and those which displace other final expenditure. Calculation of implicit SOC rate to estimate lost stream of consumption on account of displacement of other final expenditure. Discounting, at STP rate, of costs and benefits. |

| |
|---|
| Presentation of results: <br> (a) probability distributions/mean expected net present value of outcomes; <br> (b) sensitivity analysis performed with regard to key uncertain factors; <br> (c) implications for income distribution analysed including pecuniary spillovers and intra-marginal consumers' surplus. |

**Decision taker**

| |
|---|
| Reference of project to analyst. |

| |
|---|
| Selection of cut-off point for degree of remoteness of costs and benefits. |

| |
|---|
| Determination (implicitly or explicitly) of shadow prices set by policy considerations. |

| |
|---|
| Choice of STP rate to be used for discounting future consumption. |

| |
|---|
| Choice and application of decision criteria, *e.g.* <br> (a) positive mean expected present value, <br> (b) insensitive to most uncertain factors, <br> (c) decreases income inequality. |

| |
|---|
| Decision. |

20. A general model of a cost-benefit analysis. H. G. Walsh and Alan Williams, "Current Issues in Cost-Benefit Analysis," *CAS Occasional Paper, No. 11,* London: HMSO, 1969.

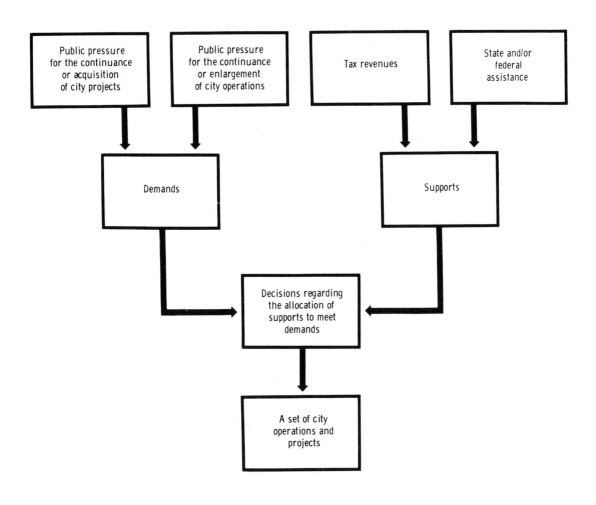

21. A functional model of the city. ABT Associates, *Applications of Systems Analysis Models,* Washington: NASA, 1968.

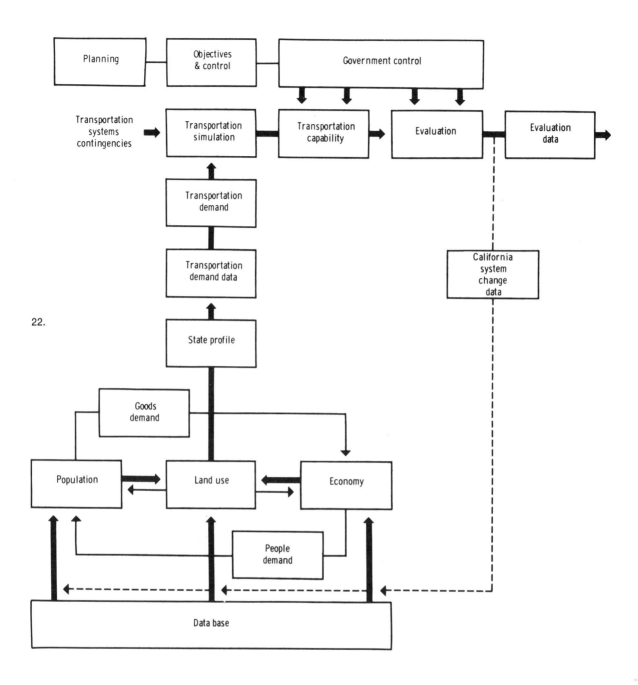

22.

22. The California transportation model. ABT Associates, *Applications of Systems.*

23. A housing model. Harry B. Wolfe and Martin Ernst, "Simulation Models and Urban Planning," in Philip M. Morse, ed., *Operations Research for Public Systems,* Cambridge: M.I.T. Press, 1967.

24. Land use and transportation model. Kenneth J. Schlager, "A Land-Use Plan Design Model," in David C. Sweet, ed., *Models of Urban Structure,* Lexington, Mass.: D.C. Heath Co., 1972.

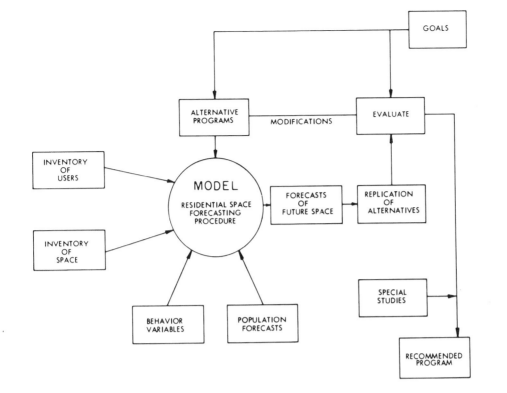

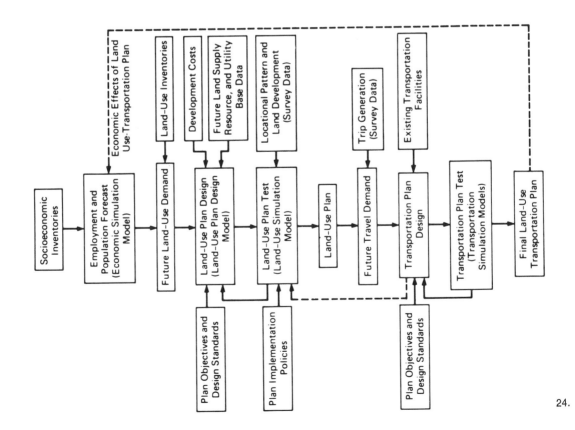

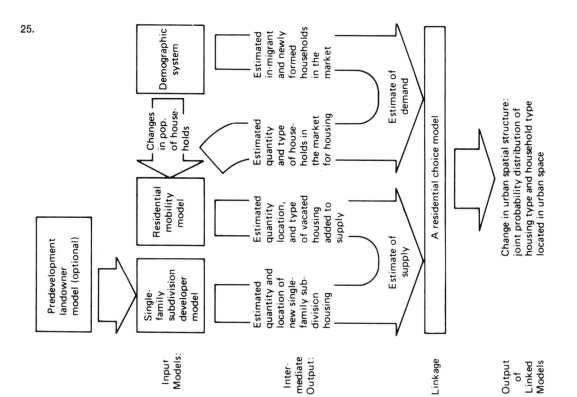

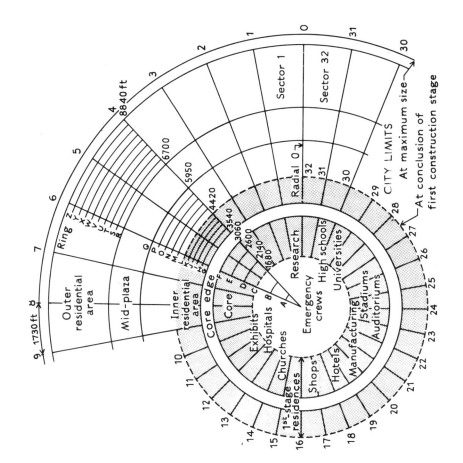

25. Linked decision model. Edward J. Kaiser, "Decision Agent Models: An Alternative Modeling Approach for Urban Residential Growth," in Sweet, ed., *Models of Urban Structure.*

26. Compact city partial plan. George B. Dantzig and Thomas L. Saaty, *Compact City,* San Francisco: W. H. Freeman Co., 1973.

27. Model of water recycling system in compact city. Dantzig and Saaty, *Compact City.*

27.

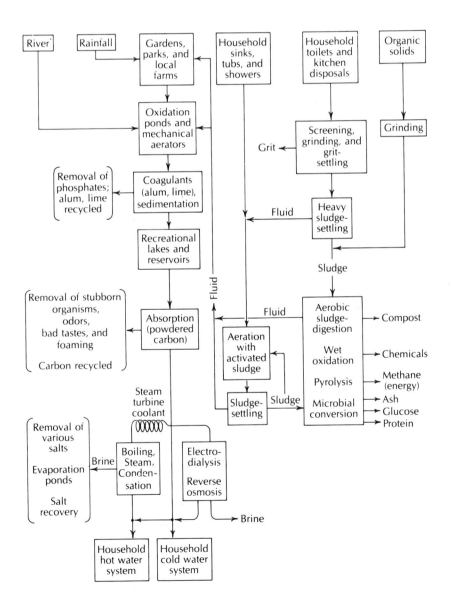

28.

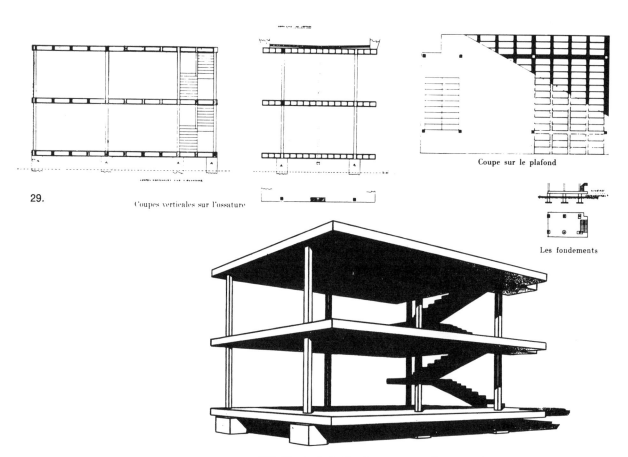

29.

Coupes verticales sur l'ossature

Coupe sur le plafond

Les fondements

L'ossature standard « Dom-ino », pour exécution en grande série

28. Sir Joseph Paxton's Crystal Palace, preparations of subassemblies at site, November 1850. Courtesy Columbia University.

29. Le Corbusier's Dom-ino system. Le Corbusier and Pierre Jeanneret, *Oeuvre Complète 1910–1929,* Zurich: Editions d'Architecture Moderne (Artémis), 1967.

30. Citrohan House by Le Corbusier. Maurice Besset, *Who Was Le Corbusier?* Geneva: Editions d'Art Albert Skira, 1968.

31. Two concrete houses at Stuttgart-Weissenhof by Le Corbusier and Pierre Jeanneret. Courtesy Columbia University.

30.

31.

32. Buckminster Fuller's Dymaxion House. Courtesy Columbia University.

33. Frank Lloyd Wright's precast panels. Courtesy Columbia University.

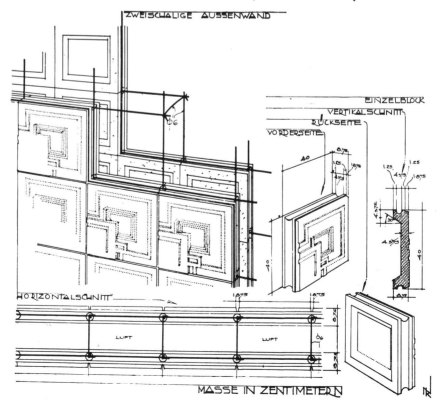

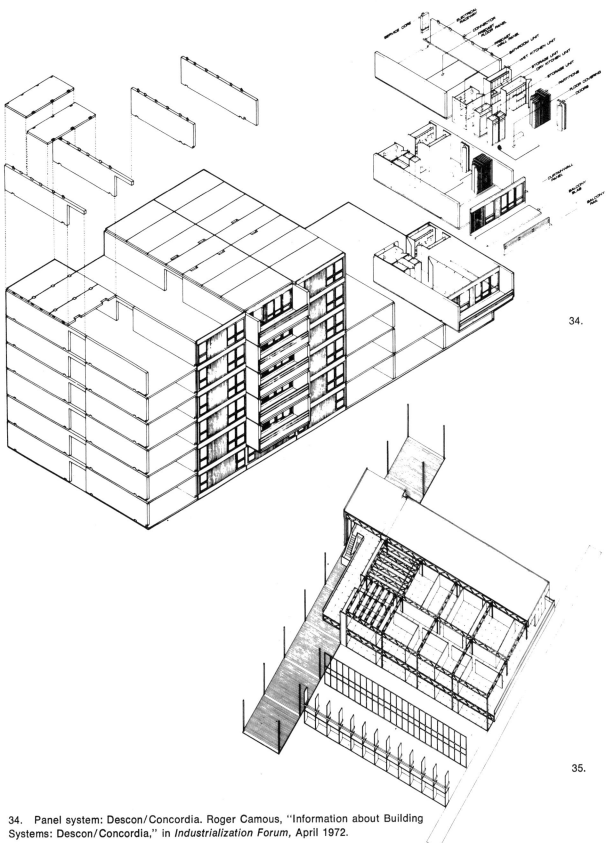

34.

35.

34. Panel system: Descon/Concordia. Roger Camous, "Information about Building Systems: Descon/Concordia," in *Industrialization Forum,* April 1972.

35. Skeletal system: Romac MODULAC. Stockton State College: Geddes, Brecher, Qualls, Cunningham Architects. *BSIC/EFL Newsletter,* June 1972.

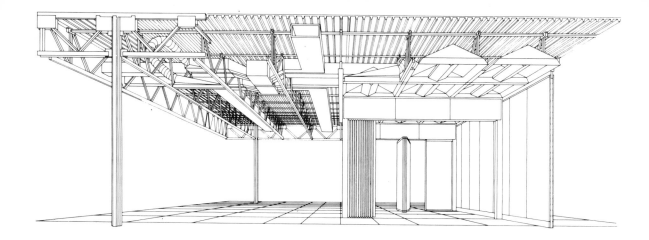

36. Skeletal system: School Construction Systems Development (SCSD). Educational Facilities Laboratories, *SCSD: The Project and the Schools,* New York: EFL, 1967.

37. Skeletal system: SCSD component heating-ventilating-air conditioning system. EFL, *SCSD: The Project and the Schools.*

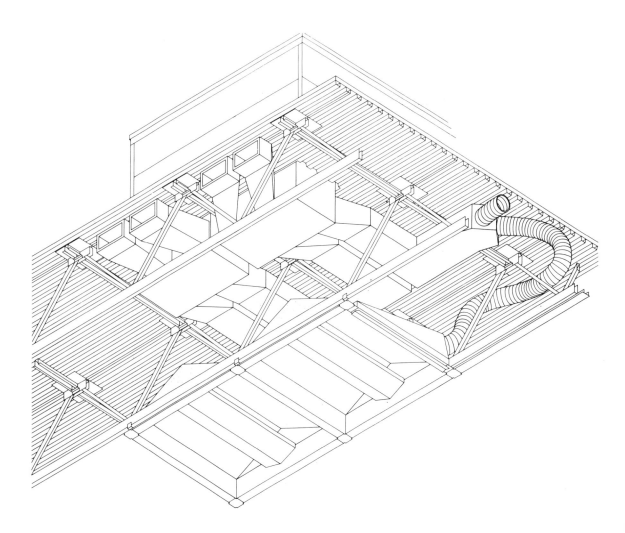

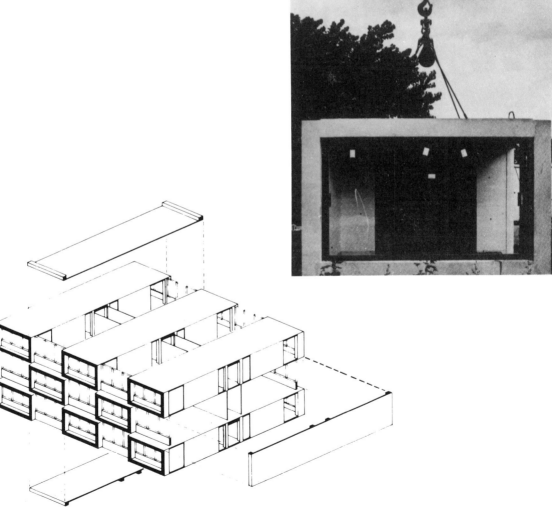

38. Cellular or box system: Shelly System. United States Department of Housing and Urban Development, *Fifth Annual H.U.D. Awards for Design Excellence*, Washington, D.C.: Department of Housing and Urban Development, 1973.

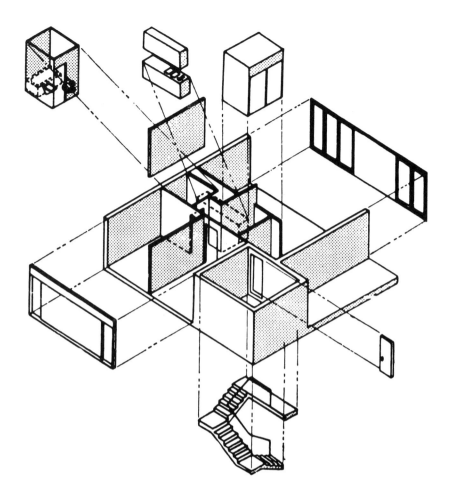

39.  Component systems. *Engineering News Record,* 183 (18), October 30, 1969.

40.  Bath component. Lawrence S. Cutler, "The Industrialized Revolution," in Albert G. H. Dietz and Lawrence S. Cutler, *Industrialized Building Systems for Housing,* Cambridge: M.I.T. Press, 1971.

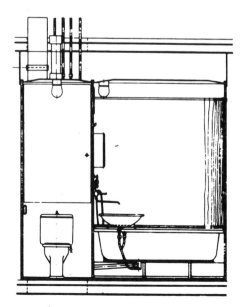

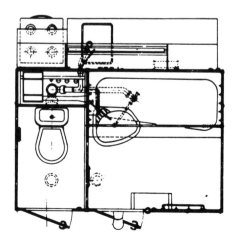

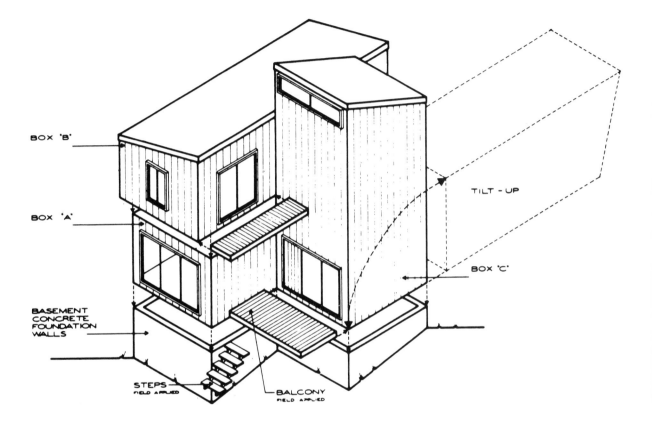

BOX 'B'

BOX 'A'

BASEMENT
CONCRETE
FOUNDATION
WALLS

STEPS
FIELD APPLIED

BALCONY
FIELD APPLIED

TILT - UP

BOX 'C'

41.   Modular system: Hercules. United States Department of Housing and Urban
Development, *Design and Development of Housing Systems for Operation Breakthrough,* Washington, D.C.: Department of Housing and Urban Development, 1973.

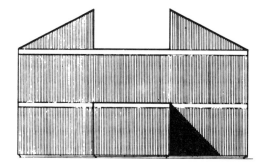

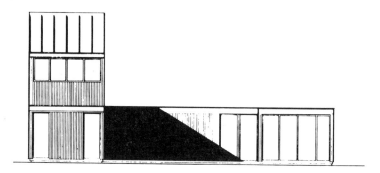

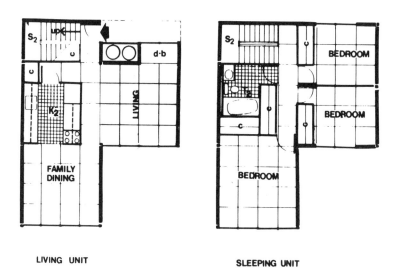

LIVING UNIT                    SLEEPING UNIT

42. Modular system: Eco-Design. Harold Finger, "A Requirement for Change: Operation Breakthrough," in Dietz and Cutler, *Industrialized Building Systems*.

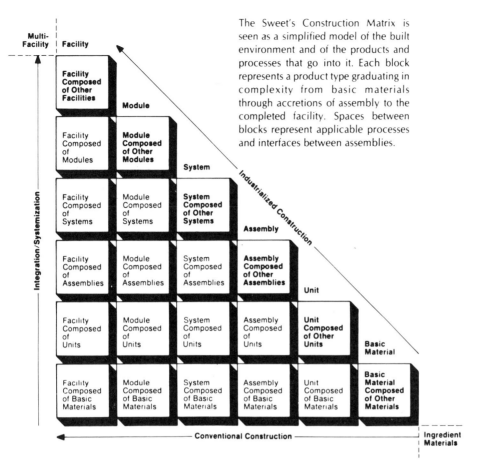

The Sweet's Construction Matrix is seen as a simplified model of the built environment and of the products and processes that go into it. Each block represents a product type graduating in complexity from basic materials through accretions of assembly to the completed facility. Spaces between blocks represent applicable processes and interfaces between assemblies.

43. Construction matrix. Miriam S. Eldar, "Sweets Refines Logic of Product Research," in *Architectural Record,* August 1973.

44. United States average hourly earnings in manufacturing and construction. Ian Donald Turner and John F. C. Terner, *Industrialized Housing,* Washington, D.C.: United States Agency for International Development, 1972.

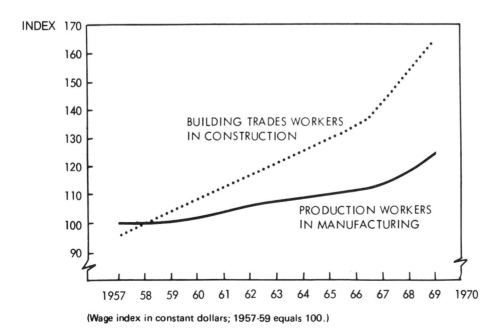

(Wage index in constant dollars; 1957-59 equals 100.)

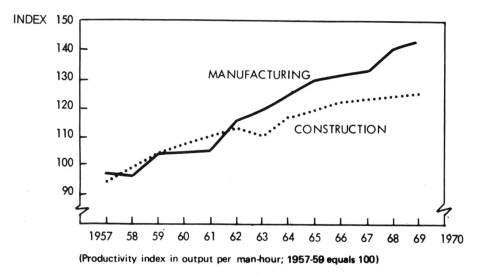

**(Productivity index in output per man-hour; 1957-59 equals 100)**

45. United States productivity in manufacturing and construction. Turner and Terner, *Industrialized Housing.*

46. Production of mobile and manufactured homes. Arthur D. Bernhardt, "The Mobile Home Industry: A Case Study in Industrialization," in Dietz and Cutler, *Industrialized Building Systems.*

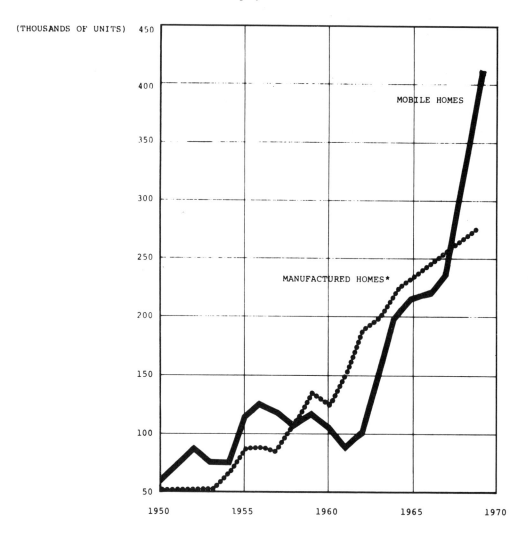

| Halldane | Cronberg et.al. | Pěna & Focke | Hattis |
|---|---|---|---|
| 1. Operational Goals<br>What is the design for? | 1. Choosing Activities relevant to function | 1. Establish Goals | 1. Problems<br>relate problems to different groups of community |
| 2. Parameters<br>What are the factors to consider in design? | 2. Defining Users and their relevant characteristics | 2. Collect, Organize and Analyze Facts | 2. Course of Action<br>mobilization of resources to solve the problems |
| 3. Synthesis<br>How are the factors related? | 3. Identifying & Structuring user requirements on the basis of step 1 & 2 | 3. Uncover and Test Programmatic Concepts | 3. Activities<br>display alternative sets of activities for each course of action |
| 4. Performance Criteria<br>What attributes and magnitudes are needed for the factors to meet the goals? | 4. Defining the Given Conditions (such as climate, law, restrictions etc.) | 4. Determine Needs | 4. Environmental Characteristics<br>quantifying environmental requirements |
| | 5. Identifying & Structuring Performance Requirements on the basis of step 3 & 4 | 5. State the Problem | |

47. Saeterdal summary of some varying procedures in identifying user needs. Anne Saeterdal, "Review of Papers," in National Bureau of Standards, *Performance Concept in Buildings: Special Publication 361*, vol. 2, Proceedings of the Joint RILEM–ASTM–CIB Symposium, held May 2–5, 1972, Philadelphia, Pennsylvania.

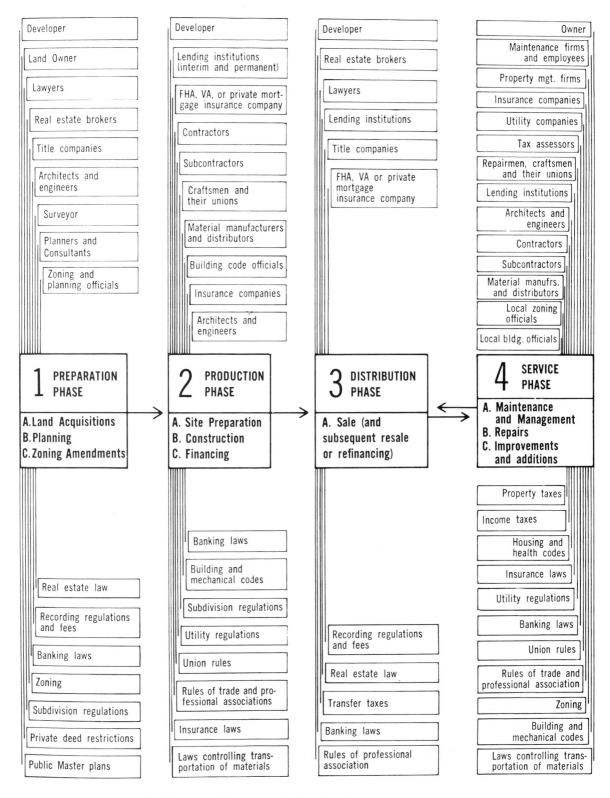

48. Institutionalization of the housing industry. The President's Committee on Urban Housing, *A Decent Home,* Washington, D.C.: The Commission, 1968.

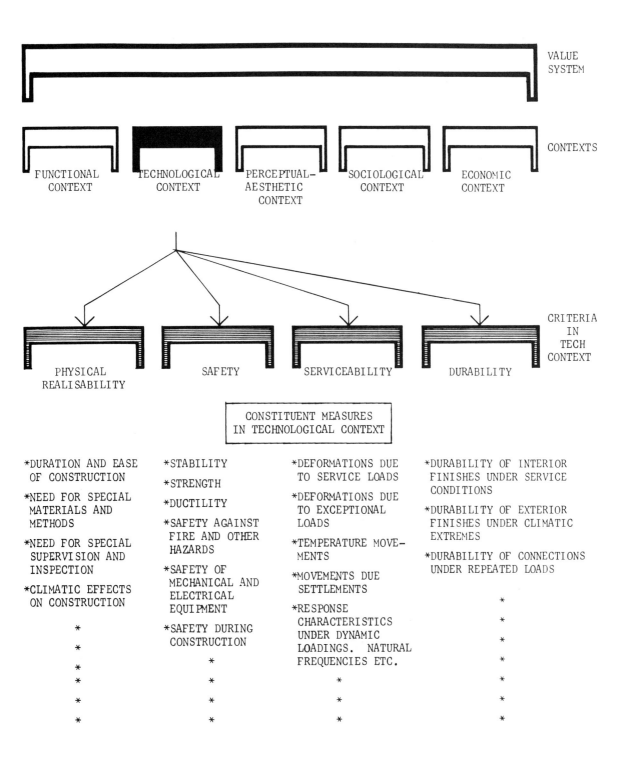

VALUE SYSTEM

CONTEXTS

FUNCTIONAL CONTEXT

TECHNOLOGICAL CONTEXT

PERCEPTUAL-AESTHETIC CONTEXT

SOCIOLOGICAL CONTEXT

ECONOMIC CONTEXT

CRITERIA IN TECH CONTEXT

PHYSICAL REALISABILITY

SAFETY

SERVICEABILITY

DURABILITY

CONSTITUENT MEASURES IN TECHNOLOGICAL CONTEXT

*DURATION AND EASE OF CONSTRUCTION

*NEED FOR SPECIAL MATERIALS AND METHODS

*NEED FOR SPECIAL SUPERVISION AND INSPECTION

*CLIMATIC EFFECTS ON CONSTRUCTION

*

*

*

*

*

*

*STABILITY

*STRENGTH

*DUCTILITY

*SAFETY AGAINST FIRE AND OTHER HAZARDS

*SAFETY OF MECHANICAL AND ELECTRICAL EQUIPMENT

*SAFETY DURING CONSTRUCTION

*

*

*

*

*DEFORMATIONS DUE TO SERVICE LOADS

*DEFORMATIONS DUE TO EXCEPTIONAL LOADS

*TEMPERATURE MOVE-MENTS

*MOVEMENTS DUE SETTLEMENTS

*RESPONSE CHARACTERISTICS UNDER DYNAMIC LOADINGS. NATURAL FREQUENCIES ETC.

*

*

*

*DURABILITY OF INTERIOR FINISHES UNDER SERVICE CONDITIONS

*DURABILITY OF EXTERIOR FINISHES UNDER CLIMATIC EXTREMES

*DURABILITY OF CONNECTIONS UNDER REPEATED LOADS

*

*

*

*

*

*

49. Values, contexts, and criteria in design assessment. Syed Gulzar Haider and Narbey Khachaturian, "A Systems Approach for the Evaluation of Performance of Buildings in Design Process," in National Bureau of Standards, *Performance Concept in Buildings.*

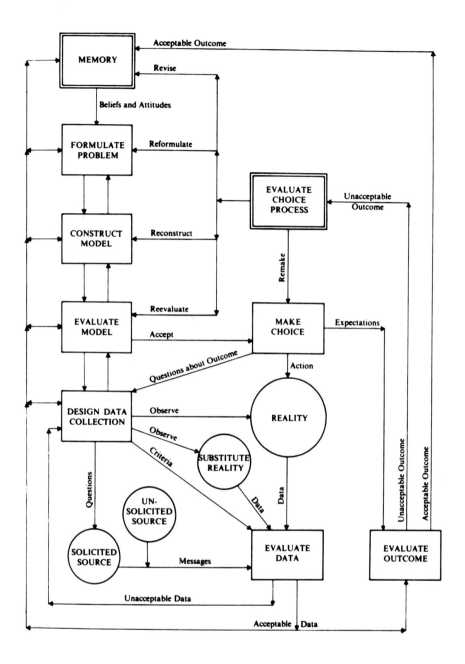

50. The choice process. Russell L. Ackoff and Fred E. Emery, *On Purposeful Systems*, Chicago: Aldine-Atherton.

51. Model of the design process. Ingvar Karlen, "Performance Concept and the System Approach," in National Bureau of Standards, *Performance Concept in Buildings*, vol. 1.

52. The building design process. Haider and Khachaturian, "A Systems Approach."

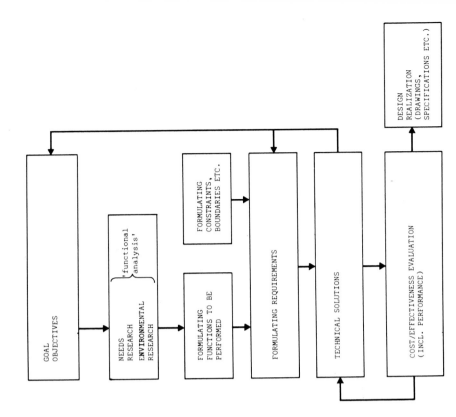

51.

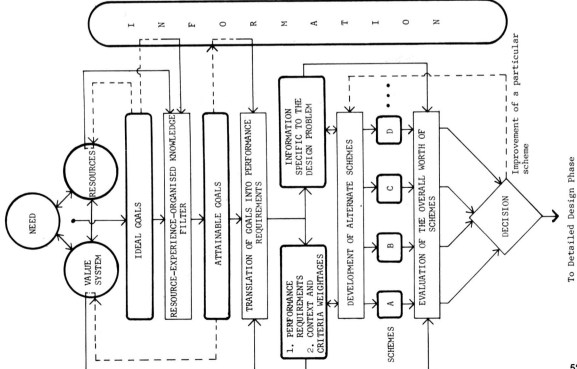

52.

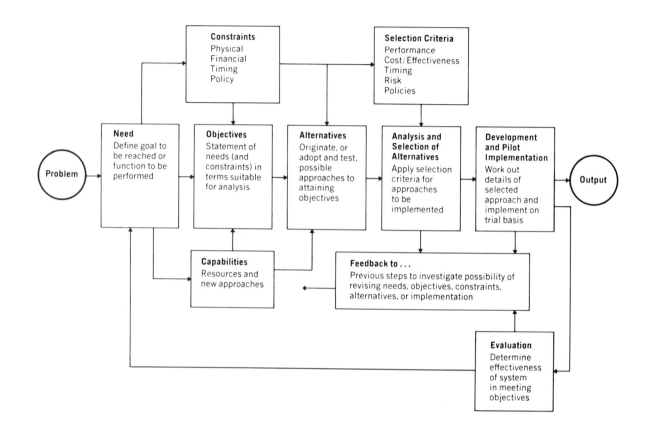

53. A systems approach at the urban scale. International City Management Association, *Applying Systems Analysis in Urban Government,* Washington, D.C.: International City Management Association for the Department of Housing and Urban Development, 1972.

54. The systems model (Fig. 53) applied to fire station location. International City Management Association, *Applying Systems Analysis.*

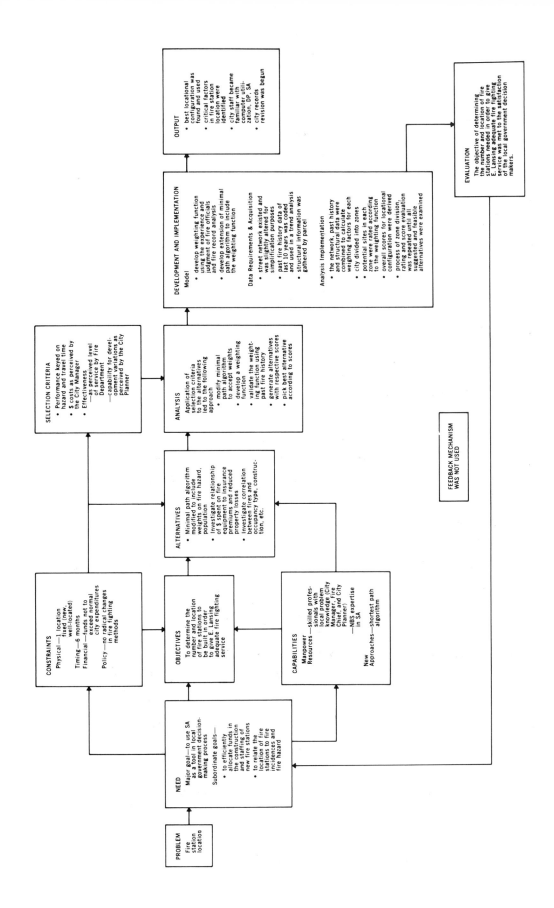

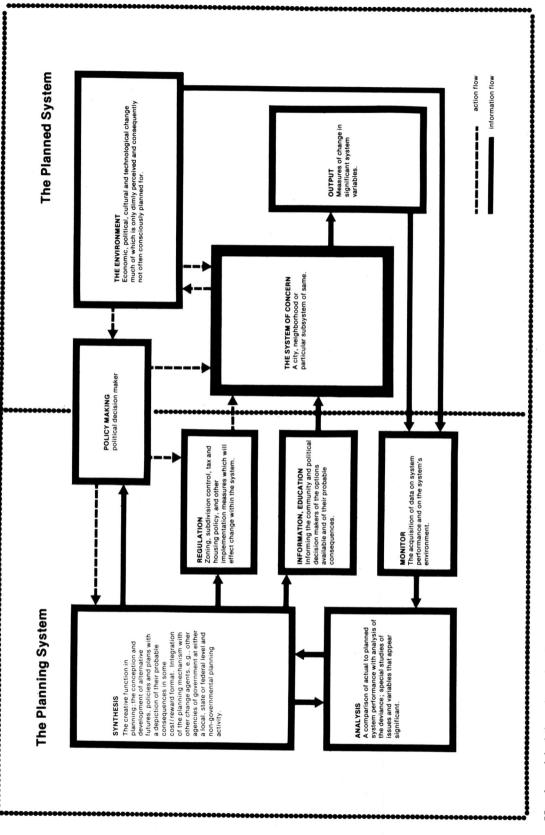

**The Planned System**

**The Planning System**

**THE ENVIRONMENT**
Economic, political, cultural and technological change much of which is only dimly perceived and consequently not often consciously planned for.

**POLICY MAKING**
political decision maker

**THE SYSTEM OF CONCERN**
A city, neighborhood or particular subsystem of same.

**OUTPUT**
Measures of change in significant system variables.

**SYNTHESIS**
The creative function in planning: the conception and development of alternative futures, policies and plans with a depiction of their probable consequences in some cost/reward format. Integration of the planning mechanism with other change agents, e.g., other agencies of government at either a local, state or federal level and non-governmental planning activity.

**REGULATION**
Zoning, subdivision control, tax and housing policy, and other implementation measures which will effect change within the system.

**INFORMATION, EDUCATION**
Informing the community and political decision makers of the options available and of their probable consequences.

**MONITOR**
The acquisition of data on system performance and on the system's environment.

**ANALYSIS**
A comparison of actual to planned system performance with analysis of the deviance; special studies of issues and variables that appear significant.

- - - - - - - action flow

━━━━━━ information flow

55. A model of the urban planning process.

# Notes

## 1. Introduction

1. An early and cogent specification of this premise which significantly influenced the recent development of design methodologies is Christopher Alexander, *Notes on the Synthesis of Form* (Cambridge: Harvard University Press, 1964).

2. Eliel Saarinen, *The City: Its Growth, Its Decay, Its Future* (New York: Reinhold, 1943), pp. 41, 44, 68, 75–79, 123.

3. Lewis Mumford, *The City in History* (New York: Harcourt, Brace and World, 1961), pp. 302–303, and Camillo Sitte, *City Planning According to Artistic Principles,* trans. by C. C. and G. R. Collins, New York, Random House, 1965, pp. 13, 41, and passim.

4. Gary T. Moore, ed., *Emerging Methods in Environmental Design and Planning* (Cambridge: MIT Press, 1970), p. viii.

5. See the "Need for New Methods," pp. 27–42 in J. Christopher Jones, *Design Methods: Seeds of Human Futures* (New York: Wiley-Interscience, 1970).

6. See "Design—A Special Kind of Decision Making," pp. 13–28 in Building Performance Research Unit, *Building Performance* (New York: Wiley, 1972) and Wojciech W. Gasparski, "The Design Activity as a Subject of Studying—The Design Methodology," in *DMG-DRS* (Design Methods Group-Design Research Society) *Journal: Design Research and Methods* 7(4): October–December 1973, pp. 306–311.

7. Explanation and justification of the planning and design process were very much in order after the Pentagon's unfortunate decision to go ahead with the ill-fated TFX-F-111 airplane. The lengthy debate on the issue offers many insights to the architect and, particularly, the urban planner. While the systems analyst's reputation for rationality suffered in the discussions, it was clearly a case of attempting incompatible objectives at the outset of the planning process, the consequences of which were not fully appreciated until the plane started to fly. See: Robert J. Art, *The TFX Decision* (Boston: Little, Brown, 1968); Committee on Government Opera-

tions, United States Senate, *TFX Contract Investigation,* Report 91–1496, 91st Congress, Washington: U.S. Government Printing Office, 1970; I. F. Stone, "Nixon and The Arms Race: The Bomber Boondoggle," in *New York Review of Books,* 2 January 1969, pp. 5–12.

8. An interesting assessment from an architectural perspective can be found in: Robert Jensen, "Buildings That Believe in Science," in *Progressive Architecture* 55(3): March 1974, pp. 82–87. See also the subsequent comment: *Progressive Architecture* 55(6): June 1974, pp. 7–8, 107.

9. Martin Kuenzlen, *Playing Urban Games: The Systems Approach to Planning* (Boston: The i Press, 1972), pp. 7–22, 40, passim.

## 2. The Systems Idea in Outline

1. "Science is built of facts the way a house is built of bricks; but an accumulation of facts is no more science than a pile of bricks is a house." Henri Poincaré, *La Science et l'hypothése* (1902).

2. Karl Mannheim, *Ideology and Utopia,* trans. Louis Wirth and Edward Shils (New York: Harvest Book, 1936), p. 276.

3. The broader philosophical use of the systems idea is not accounted for in this essay. A basic introduction is provided in Ervin Laszlo, *The Systems View of the World* (New York: Braziller, 1972). The most prominent current advocates of the concept are general systems theorists. Their perspective is succinctly summarized by O. R. Young ("A Survey of General Systems Theory," *General Systems Yearbook,* 9, 1964: pp. 61–80, p. 61).

> The central unifying concept which they came up with was the notion of a system. A system in this instance may be somewhat loosely defined as a set of objects together with relationships between the objects and between their attributes. It was felt that at a basic level all of the disciplines concerned must deal with systems of one kind or another and that there must be a goodly number of basic orienting concepts which are relevant to systems of all kinds. Out of this feeling grew the concept of isomorphism and isomorphies which suggests that at a basic level there exists a body of theoretical principles which can be applied usefully to systems of all kinds and from all disciplines. In a sense then the subsequent history of general systems theory traces a record of events to elaborate the basic principles of a general system and to apply these principles to the specific and concrete systems of interest to various fields of study.
>
> It is obvious that one of the main bases of general systems theory is a belief in the usefulness of a synthetic approach combining contributions from many fields in order to achieve advances which can in turn be communicated back to specific fields. It is therefore interesting to note that the work done to date in the area of general systems is of a highly synthetic nature involving contributions from a number of separate disciplines in both the natural and social sciences. The reverse flow of the concepts of general systems theory to applications in the specific disciplines does not, however, appear to cover quite as broad a spectrum at this time.

This expression of the idea has roots in philosophic extensions of biology. See, for example: Joseph Needham, *Order and Life* (New Haven: Yale University Press, 1936), pp. 249–250, 263–265; E. H. Sutherland, "Biological and Sociological Processes," in Ernest W. Burgess, ed., *The Urban Community* (Chicago: University of

Chicago Press, 1926), p. 70; Ludwig von Bertalanffy, *Problems of Life* (New York: Wiley, 1952), pp. 11–12. For an introduction to general systems theory, see: Kenneth E. Boulding, "General Systems Theory: The Skeleton of Science," *Management Science* 2(3): April 1956, pp. 197–208; *The Image* (Ann Arbor: University of Michigan Press, 1961), and V. A. Lektorsky and V. N. Sadovsky, "On Principles of System Research," *General Systems Yearbook,* 5: 1960, pp. 171–179; W. Ross Ashby, "General Systems as a New Discipline," *General Systems Yearbook* 10, 1965; Ludwig von Bertalanffy, "General Systems Theory: A Critical Review," *General Systems Yearbook* 7, 1962. Perhaps the most prominent contemporary philosopher with a systems or organic perspective was Alfred North Whitehead. See: Joseph Needham, "A Biologist's View of Whitehead," Paul A. Schilpp, ed., *The Philosophy of Alfred North Whitehead* (New York: Tudor, 1951, 2nd edition); Dorothy M. Emmet, *Whitehead's Philosophy of Organism* (London: Macmillan, 1932); Alfred North Whitehead, *Process and Reality* (New York: Macmillan, 1929); Felix Cesselin, *La Philosophie Organique de Whitehead* (Paris: Presses Universitaires de France, 1950).

4. Edward S. Quade, *Cost-Effectiveness: An Introduction and Overview* (Santa Monica: Rand, 1965). This relates to C. P. Snow's definition of science as nothing but organized common sense. Sir Charles Percy Snow, *Science and Government* (Cambridge, Mass.: Harvard University Press, 1961). Rudolf Klein, defines systems analysis as: ". . . constructing a highly simplified model of the 'real' world and expressing in a series of mathematical equations the relationship among the factors selected," in "Growth and Its Enemies," *Commentary,* 53(6): June 1972, pp. 37–44, p. 37.

5. Chris Argyris, "Understanding Human Behavior in Organization," in Mason Haire, ed., *Modern Organizational Theory* (New York: Wiley, 1959), pp. 124–125.

6. Charles J. Hitch and Roland McKean, *The Economics of Defense in The Nuclear Age* (New York: Atheneum, 1967), pp. 128–130.

7. For a more beneficent vision of this planning, see: Robert Moses, *Public Works: A Dangerous Trade* (New York: McGraw-Hill, 1970), pp. 301–309. Also see Robert Caro, *The Power Broker; Robert Moses and the Fall of New York* (New York: Knopf, 1974).

8. Earl A. Levin, "Business, Balkanism and Blindness," *Planning 1965* (papers of a joint planning conference of the American Society of Planning Officials and the Community Planning Association of Canada; Chicago: American Society of Planning Officials, 1965), p. 130.

9. Amitai Etzioni, "Mixed Scanning: A 'Third' Approach to Decision Making," *Public Administration Review* 27(5): December 1967, pp. 385–392, p. 385.

10. Albert O. Hirschman and Charles E. Lindblom, "Economic Development, Research and Development, Policy Making: Some Converging Views," *Behavioral Science* 7(2): April 1962, p. 215.

11. Gui Bonsiepe, "Arabesques on Rationality," *Ulm,* 19/20: August 1967, pp. 9–23, p. 12.

12. Edward C. Banfield, "Ends and Means in Planning," in Sidney Mailick and Edward H. Van Ness, eds., *Concepts and Issues in Administrative Behavior* (Englewood Cliffs: Prentice-Hall, 1962), p. 71.

13. David Braybrooke and Charles E. Lindblom, *A Strategy of Decision* (New York: The Free Press, 1963), p. 9.

14. Richard S. Bolan, "Emerging Views of Planning," *Journal of the American Institute of Planners* 33(4): July 1967, p. 241.

15. For an introduction to these techniques, see: Richard I. Levin and Rudolph P. Lamone, *Quantitative Disciplines in Management Decisions* (Belmont, California: Dickenson Publishing Co., 1969); George Chadwick, *A Systems View of Planning* (New York: Pergamon Press, 1971); Ira M. Robinson, ed., *Decision-Making in Urban Planning* (Beverly Hills: Sage Publications, 1972); David C. Sweet, *Models of Urban Structure* (Lexington, Mass.: D. C. Heath, 1972).

16. Frank H. Knight, *Freedom and Reform* (New York: Harper, 1947), p. 344; for some other views on this, see Harold Garfinkle, "The Rational Properties of Scientific and Common Sense Activities," *Behavioral Science* 5(1): January 1960, pp. 72–83, and John M. Pfiffner, "Administrative Rationality," *Public Administration Review* 20(1): Winter 1960, pp. 125–32.

17. John Dewey, *How We Think* (Boston: Heath, 1910), p. 12.

18. Ibid., p. 72 for Dewey's version of this. Also, see Sir Francis Bacon, *Of the Proficience and Advancement of Learning, Divine and Humane* (New York: Da Capo, Repr. of 1605 edition).

19. Charles J. Hitch and Roland McKean, *Economics of Defense.*

20. Robert N. Lehrer, *The Management of Improvement* (New York: Reinhold, 1965), p. 115; see also Roland N. McKean, *Efficiency in Government Through Systems Analysis* (New York: Wiley, 1958).

21. Charles J. Hitch and Roland McKean, *Economics of Defense,* pp. 118–119.

22. Herbert A. Simon, "Architecture of Complexity," *General Systems Yearbook* 3, 1965, p. 76. See also: Herbert A. Simon, "The Logic of Heuristic Decision Making," in N. Rescher, ed., *Logic of Decision and Action* (Pittsburgh: Pittsburgh University Press, 1967).

## 3.  Holism: Conceptions of System

1. *Oxford Universal Dictionary,* 3rd Edition, 1955, p. 2115.

2. Edward V. Krick, *An Introduction to Engineering and Engineering Design* (New York: Wiley, 1965), p. 211.

3. A. D. Hall and R. E. Fagen, "Definition of System," in Walter Buckley, ed., *Modern Systems Research for the Behavioral Scientist* (Chicago: Aldine Publishing Co., 1968), p. 81.

4. Ludwig von Bertalanffy, "General Systems Theory: A Critical Review," in *General Systems Yearbook* 7, 1962.

5. Colin Cherry, *On Human Communication* (New York: M.I.T. Press, 2nd ed. 1968), p. 307.

6. Allen R. Stubberud, *Analysis and Synthesis of Linear Time-Variable Systems* (Berkeley: University of California Press, 1964), p. 1.

7. Krick, *Engineering Design,* pp. 211–212.

8. Robert N. Lehrer, *The Management of Improvement* (New York: Reinhold, 1965), p. 115.

9. Robert A. Dahl, *Modern Political Analysis* (Englewood Cliffs, New Jersey: Prentice Hall, 1963), p. 9.

10. Roy R. Grinker, ed., *Toward a Unified Theory of Human Behavior* (New York: Basic Books, 1956), p. 370. See also Kingsley Davis, *Human Society* (New York: Macmillan, 1949), pp. ix, 9, 24–51.

11. Robert A. Dahl, *Modern Political Analysis,* p. 10.

12. James G. Miller, "Toward a General Theory for the Behavioral Sciences," in Leonard D. White, ed., *The State of the Social Sciences* (Chicago: The University of Chicago Press, 1956), pp. 29–65.

13. Alexander H. Leighton, *My Name Is Legion: Foundations for a Theory of Man in Relation to Culture* (New York: Basic Books, 1959): volume I of the *Stirling County Study of Psychiatric Disorder and Socio-cultural Environment;* Charles Hughes, Marc-Abelard Tremblay, Robert N. Ranport, Alexander H. Leighton, *People of Cove and Woodlot* (New York: Basic Books, 1960), volume II of the Stirling County Study; Dorothea Leighton, John S. Harding, David B. Macklin, Allister M. Macmillan, Alexander H. Leighton, *The Character of Danger* (New York: Basic Books, 1963), volume III of the Stirling County Study.

14. Leighton, *My Name Is Legion,* p. 197.

15. Ibid., p. 199.

16. Ibid., p. 200. For a more detailed view of a particular city (San Francisco) whose systemic properties are perceived through the relationships of a specific physical sub-system (housing), see: Cyril Herrmann, "The City as a System" in National Academy of Sciences and The National Academy of Engineering, *Science, Engineering and The City* (Washington: National Academy of Sciences, 1967), pp. 118–131.

17. Leighton, *My Name Is Legion,* p. 196; see also p. 424.

18. Dahl, *Modern Political Analysis,* p. 9.

19. Herbert A. Simon, "The Architecture of Complexity," in *General Systems Yearbook,* 10, 1965, p. 63.

20. Kenneth L. Kraemer, *Policy Analysis in Local Government: A Systems Approach to Decision-Making* (Washington: International City Management Association, 1973), p. 31.

## 4. Rationality: Manifestations and Instruments

1. John Weightman, "Cultivating the Enlightenment," *New York Review of Books* 7(12): January 1967, pp. 4, 12. For a further look at the difficulty in dealing with valuation in planning, see "On the Objectivity of Social Science," in Richard S. Rudner, *Philosophy of Social Science* (Englewood Cliffs: Prentice-Hall, 1966), pp. 68–83; "The Subjective Character of the Data of the Social Sciences," in Friedrich A. Hayek, *The Counter Revolution of Science* (Glencoe: Free Press, 1952), pp. 25–35; "Basic Questions of Values," in Alfred De Grazia, *Politics and Government* (New York: Collier Books, 1962), vol. 1, pp. 303–307; "Fact and Value in Decision Making," in Herbert A. Simon, *Administrative Behavior* (New York: Free Press, 1965); "A Critique of the Classical Ideals," in David Braybrooke and Charles E. Lindblom, *A Strategy Of Decision* (New York: The Free Press of Glencoe, 1963); Wayne A. R. Leys, "The Value Framework of Decision Making," in Mailick and Van Ness, *Concepts and Issues in Administrative Behavior* (Englewood Cliffs: Prentice-Hall, 1962); Gunner Myrdal, *Value in Social Theory* (London: Routledge and Kegan Paul, 1958); Charles E. Lindblom, *The Intelligence of Democracy* (New York: Free Press, 1965); and "Concerning Political Ends," in George E. G. Catlin, *Systematic Politics* (Toronto, Canada: University of Toronto Press, 1962), pp. 385–423.

2. See: C. W. Cassinelli, "The Public Interest in Political Ethics"; Gerhart Niemeyer, "Public Interest and Private Utility"; Harold D. Lasswell, "The Public Inter-

est: Proposing Principles of Content and Procedure": all in Carl J. Friedrich, *The Public Interest* (New York: Atherton, 1966), pp. 1–13, 44–79.

3. Paul Diesing, *Reason in Society* (Urbana: University of Illinois Press, 1962), p. 1.

4. Ibid., p. 20.

5. Ibid., pp. 14, 17, 43.

6. Ibid., p. 170.

7. Creation of totally new environments, i.e., a building, new town, etc., being a special case: a modification of the existing natural condition.

8. A very influential formulation of the problem-solving orientation is Alan Newell, J. C. Shaw, and Herbert A. Simon, "Elements of a Theory of Human Problem Solving," *Psychological Review,* volume 65 (1958). Since their perception of the design process may raise the hackles of the more aesthetically oriented, it must be stressed that design can be viewed legitimately as a problem-solving process. See, for example, David W. Ecker, "The Artistic Process as Qualitative Problem Solving," *Journal of Aesthetics and Art Criticism,* 21(3): Spring 1963, pp. 283–290.

9. Edward V. Krick, *An Introduction to Engineering and Engineering Design* (New York: Wiley, 1965), p. 3.

10. Ibid., pp. 4–5. This is obviously a simplification of the definition of a problem. How one perceives a problem involves a complex network of activities which are reviewed in our discussion of "Abstraction and Model Building" that follows.

11. Dimitrius N. Chorafas, *Systems and Simulation* (New York: Academic Press, 1965), p. 1.

12. Anatol Rapoport, *Two Person Game Theory* (Ann Arbor: University of Michigan Press, 1956), pp. 5–6.

13. John Dewey, *How We Think* (Boston: Heath, 1910), p. 72.

14. George A. Steiner, *Managerial Long-Range Planning* (New York: McGraw-Hill, 1963), p. 26.

15. Braybrooke and Lindblom, *A Strategy of Decision,* pp. 10 and 41, and Garfinkle, *The Rational Properties,* p. 76. There is, of course, an easy dismissal of attempts at rationality in solving social problems by those with a "hippie" or nihilist viewpoint. These positions are both too numerous and trivial to account for here. But what is of concern is the uneasiness of those who do look for rational solutions to social problems. Some of the views which I have found most persuasive and which have influenced the thinking in this essay include: Daniel P. Moynihan, *Coping* (New York: Random House, 1974); Robert L. Heilbroner, *An Inquiry into the Human Prospect* (New York: Norton, 1974); René Dubos, *A God Within* (New York: Scribners', 1972); Walter A. Weisskopf, *Alienation and Economics* (New York: Dutton, 1971); Hannah Arendt, "Washington's 'Problem-Solvers'—Where They Went Wrong," *New York Times,* 5 April 1972; Hannah Arendt, *Crises of the Republic* (New York: Harcourt Brace Jovanovich, 1972); Nathan Glazer, "The Limits of Social Policy," *Commentary* 52(3): September 1971, pp. 51–58; Jeffrey L. Pressman and Aaron B. Wildavsky, *Implementation* (Berkeley: University of California Press, 1973).

16. Martin Shubik, "Studies and Theories of Decision Making," *Administrative Science Quarterly* (3), December 1958, p. 298; and James G. March and Herbert A. Simon, *Organizations* (New York: Wiley, 1958), p. 137.

17. The Hitch-McKean "classic" construct of the systems model. Ida Hoos gives us a short version of it: ". . . systems analysis is a method of analyzing a problem by: (1) defining it; (2) ascertaining the desired objectives; (3) assessing the avail-

able resources; and (4) discovering the optimal alternatives available for achieving the objectives." In Ida R. Hoos, *Systems Analysis in State Government* (Berkeley, Calif.: Space Sciences Laboratory, University of California, Working Paper No. 79, December 1967), p. 2; for an economic expression of the idea, see: Shubik, "Studies and Theories," pp. 292, 297; the model may also be only a normative ideal: see Jacob Marschak, "Rational Behavior, Uncertain Prospects and Measurable Utility," *Econometrica* 18(2): April 1950, p. 111.

18. Irwin D. J. Bross, *Design For Decision* (New York: The Free Press, 1953), pp. 19–20.

19. Harold Koontz and C. J. O'Donnell, *Principles of Management,* 3rd ed. (New York: McGraw-Hill, 1964), pp. 81–85. For other versions of the process model see: Steiner, *Managerial Long-Range Planning,* pp. 318–319; Banfield, "Ends and Means in Planning," in Mailick and Van Ness, eds., *Concepts and Issues in Administrative Behavior* (Englewood Cliffs: Prentice-Hall, 1962); and Martin Meyerson and Edward C. Banfield, *Politics, Planning and the Public Interest* (Glencoe: The Free Press, 1955), p. 22. Even the Army uses a variant of this process, see: U.S. Army, Staff Officer's Field Manual, FM 101–5, 1960, p. 142.

20. United States Air Force Systems Command, *AFSCM 375–1: Configuration Management During Definition and Acquisition Phases* (Washington: Government Printing Office, 1964); United States Air Force Systems Command, *AFSCM 375–3: System Program Office Manual* (Washington: Government Printing Office, 1964); United States Air Force Systems Command, *AFSCM 375–4: System Program Management Procedures* (Washington: Government Printing Office, 1966); United States Air Force Systems Command, *AFSCM 375–5: Systems Engineering Management Procedures* (Washington: Government Printing Office, 1966). The Navy also has highly developed techniques for managing the planning and design process. See, for example, Department of the Navy, Headquarters Naval Materiel Command, *Configuration Management: A Policy and Guidance Manual* (Washington: Naval Materiel Command, Document NAVMATINST 4130.1, 1967).

21. *AFSCM 375–4,* p. 1.

22. Ibid., p. 6

23. Ibid., p. 8.

24. Ibid., p. 69.

25. *AFSCM 375–5,* p. 1.

26. Ibid., p. 7.

27. Ibid., p. 1.

28. Ibid., p. 7.

29. Ibid., p. 8.

30. Ibid., p. 19.

31. Ibid., pp. 19–20.

32. Ibid., pp. 4, 87–88.

33. For an introduction to this subject, see: Richard Layard, ed., *Cost-Benefit Analysis* (Baltimore: Penguin Books, 1972), and E. J. Mishan, *Elements of Cost-Benefit Analysis* (London: George Allen and Unwin, 1972).

34. Perceptive criticism of this technique can be found in: Armen A. Alchian, "Cost Effectiveness of Cost Effectiveness," in Stephen Enke, ed., *Defense Management* (Englewood Cliffs, N.J.: Prentice-Hall, 1967), pp. 74–86; Francis X. Splane, *The General Inapplicability of Cost-Effectiveness Analysis.* Paper presented to the Annual Meeting of the Pennsylvania Statistical Association, Middletown, Pa., 16 April 1971 (mimeo).

35. *AFSCM 375-5,* see pp. 94–95 for an explanation and Figure 22, p. 131 for an example of such a study.

36. Paul Lester Wiener and Francis Ferguson, *The Development Plan of New Providence Island and the City of Nassau* (London: Government of the Bahama Islands, 1968).


## 5. Architectural and Planning Antecedents of the Systems Idea

1. Perhaps the idea originated even earlier; see Howard Becker and Harry Elmer Barnes, *Social Thought from Lore to Science,* 3 vols. (New York: Dover, 1961), vol. 1, pp. 80, 127. The best introduction to the early "organic" conception of the systems idea is G. N. Giordano Orsini, "The Ancient Roots of a Modern Idea," in G. S. Rousseau, ed., *Organic Form: The Life of an Idea* (London: Routledge and Kegan Paul, 1972), pp. 7–23; In Plato, the most explicit expression is *Phaedrus,* 264c. In Aristotle: *Poetics* VIII, 51a; *Politics,* I, 2, 1253a; *Metaphysics* V, 11, 1019a; *Politics and the Athenian Constitution,* 1253a, 1328b, 1290b. See also John Warrington, "Translator's Introduction" to *Politics and the Athenian Constitution* (New York: Dutton, 1959), p. xi. Plato's use of the concept is analyzed in Ernest Barker, *Greek Political Theory: Plato and His Predecessors* (New York: Barnes and Noble, 1960), pp. 270–271. For some slightly variant interpretations see: Walter Beach, *The Growth of Social Thought* (New York: Charles Scribner, 1939), pp. 35–38, 41–42, 44–50; Rollin Chambliss, *Social Thought: From Hammurabi to Comte* (New York: Holt, Rinehart and Winston, 1954), pp. 166–168; Charles A. Ellwood, *A History of Social Philosophy* (New York: Prentice-Hall, 1938), pp. 22–23, 36–37, 40, 43; Paul H. Furfey, *A History of Social Thought* (New York: Macmillan, 1942), pp. 120, 178. Becker and Barnes, *Social Thought,* vol. II, p. 678; Emory S. Bogardus, *A History of Social Thought* (Los Angeles: Miller, 1928), p. 137. Aristotle's conception again became prominent in the medieval Thomistic synthesis, see: Otto Gierke, *Political Thought of the Middle Ages,* trans. F. W. Maitland (Cambridge, England: The University Press, 1958), pp. 7, 8, 23, 131, 132; E. Lewis, "Organic Tendencies in Medieval Political Thought," *American Political Science Review,* 32 (1938), pp. 849–876; A. H. Chroust, "The Corporate Idea and the Body Politic in the Middle Ages," *Review of Politics,* 9 (1947), pp. 423–452; Otto Gierke, *Natural Law and the Theory of Society,* trans. E. Barker, 2 vols. (Cambridge, England: The University Press, 1934).

2. See the preface to Stephen Enke, *Defense Management* (Englewood Cliffs, N.J.: Prentice-Hall, 1967), pp. v–x; Robert H. Roy, "The Development and Future of Operations Research and Systems Engineering" in Charles D. Flagle, William H. Huggins and Robert H. Roy, *Operations Research and Systems Engineering* (Baltimore: John Hopkins, 1960), pp. 8–27.

3. Russell L. Ackoff and Fred E. Emery, *On Purposeful Systems* (New York: Aldine-Atherton, 1972), p. vii.

4. Ludwig von Bertalanffy, *General System Theory* (New York: Braziller, 1968), pp. 10–29.

5. The relationship between sociology and organic-systems thought in philosophy and biology has often been a hazy one, see: E. B. Reuter, "Sociology and Biology," in Ernest W. Burgess, ed., *The Urban Community* (Chicago: University of Chicago Press, 1926), p. 67. Perhaps the area of their most significant recent interaction has been in human ecology, an "effort to employ biological concepts in the

explanation of human behavior." James M. Beshers, *Urban Social Structure* (New York: The Free Press, 1962), p. 67. In another context, it was the outcome of a concern with "excessive analysis" in viewing communities with little attempt at the synthesis of data which would provide a view of communities as coherent wholes. E. A. Gutkind, *Community and Environment* (London: Watts, 1953), pp. xii, 4. See also: Robert E. Park, "Modern Society," in Robert Redfield, ed., *Levels of Integration in Biological and Social Systems* (Lancaster, Penn.: Jacques Cattell Press, 1942), pp. 218-219; also Park's "The Urban Community as a Spatial Pattern and a Moral Order," in Burgess, ed., *The Urban Community,* p. 3; Alfred E. Emerson, "Basic Comparisons of Human and Insect Societies," in Redfield, ed., *Levels of Integration,* pp. 166-167; Hawley, *Human Ecology,* pp. 8-9, 207-210; Ian Whitaker, "The Nature and Value of Functionalism in Sociology," in Don Martindale, ed., *Functionalism in the Social Sciences* (Philadelphia: American Academy of Political and Social Science, 1965), p. 127; W. G. Runciman, *Social Science and Political Theory,* pp. 3, 110; Ernest Nagel, "Problems of Concept and Theory Formation in the Social Sciences," in Natanson, ed., *Philosophy of the Social Sciences,* p. 191; Bogardus, *A History of Social Thought,* p. 310; George C. Homans, *The Human Group* (New York: Harcourt, Brace and World, 1950), p. 268; Martindale, ed., *Functionalism in the Social Sciences,* pp. ix, 144; see also Kingsley Davis, "The Myth of Functional Analysis as a Special Method in Sociology and Anthropology," *American Sociological Review,* 24, 1959; Robert Redfield, "Societies and Cultures as Natural Systems," in Margaret Park Redfield, ed., *Human Nature and the Study of Society: Papers of Robert Redfield* (2 vols.; Chicago: University of Chicago Press, 1962), I, p. 122; I. C. Jarvie, "Limits of Functionalism and Alternatives to It in Anthropology," in Martindale, ed., *Functionalism in the Social Sciences,* p. 19; Robert K. Merton, *Social Theory and Social Structure* (Glencoe: The Free Press, 1957), pp. 74-78; Spencer, *The Nature and Value of Functionalism in Anthropology,* p. 1; Radcliffe-Brown, *Structure and Function* (Glencoe: The Free Press, 1952), pp. 12, 13; Martindale, *Sociological Theory,* p. 92; Robert Redfield, *The Little Community* (Upsala: Almquist and Wiksells, 1955), pp. 15-16, 35; Redfield, "Societies and Cultures," pp. 124-125.

6. Ernest Barker saw the contemporary resuscitation of Plato's organicism in viewing society beginning with Rousseau: Barker, *Greek Political Theory,* pp. 452-453. Others, however, see the modern regeneration of the concept being initiated with Montesquieu, e.g., Max Lerner, "Social Process," in R. A. Seligman, ed., *Encyclopedia of the Social Sciences* (New York: Macmillan, 1931), vol. 14, pp. 148-151; and A. R. Radcliffe-Brown, *Structure and Function in Primitive Society,* p. 5. But given the pervasiveness of the idea in so many realms of thought, it is difficult to imagine any one individual as being responsible for its reintroduction; cf. F. W. Coker, *"Organismic Theories of the State"* (Columbia University, 1910), pp. 9-16. Don Martindale goes even further in saying that all philosophic idealism is reflected in various extensions of the organic analogy. As a result, the very first theoretical constructions in sociology were organic ones: Don Martindale, *The Nature and Types of Sociological Theory* (Boston: Houghton Mifflin, 1960), pp. 52-56, 62, 65, 69, 78-80, 83, 86; Pitirim A. Sorokin, *Society, Culture and Personality: Their Structure and Dynamicism* (New York: Cooper Square Publishers, 1962), pp. 23, 24, 150; Sorokin, *Contemporary Sociological Theories* (New York: Harper and Bros., 1928), pp. 195-197; Becker and Barnes, *Social Thought,* II, pp. 677-692; Bogardus, *Development of Social Thought* (New York: Longmans, Green and Co., 1961, 4th edition), pp. 295-299; Ellwood, *A History of Social Philosophy,* pp. 467-478. Comte, Auguste, *The*

*Positive Philosophy* (London: Chapman, 1853), p. 27. Tonnies, *Community and Society*, p. 35, 105; Herbert Spencer was the dominant organicist, see: Herbert Spencer, *Principles of Sociology* (New York: Appleton and Co., 1896) 2 vols., I, pp. 449–462, 469, 491–548, 590–592; Becker and Barnes, *Social Thought*, pp. 680, 681. Beach, *Growth of Social Thought*, pp. 121, 179; Emory S. Bogardus, *The Development of Social Thought*, pp. 293–295; Ellwood, *History of Social Philosophy*, p. 455; Jay Rumney, *Herbert Spencer's Sociology* (New York: Atherton Press, 1966).

7. See, for example, Katherine Gilbert, "Clean and Organic: A Study in Architectural Semantics," *Journal of the Society of Architectural Historians* 10(3): October 1951, pp. 3–7. The clearest definition of the organic concept in art is G. N. Giordano Orsini's, in *Dictionary of the History of Ideas,* Philip P. Wiener, ed. (New York: Scribners, 1973), vol. 3, p. 421:

> Aesthetic organicism usually refers to the doctrine of organic unity and to its cognates like the idea of organic form or of "inner" form. The designation arises from the assumption that a work of art may be compared to a living organism, so that the relation between the parts of a work is neither arbitrary nor factitious, but as close and intimate as that between the organs of a living body. The classic formula for this relation is double: (1) the parts of the work are in keeping with each other and with the whole, and (2) alteration of a part will bring with it the alteration of the whole. By means of this formula the closest unity between the parts of a work of art is predicated or, alternatively, the formula provides the closest way of conceiving aesthetic unity.

For an elaboration of the concept's implications see: Stephen C. Pepper, "Organistic Criticism," in *The Basis for Criticism in the Arts* (Cambridge: Harvard, 1963), pp. 74–95; Meyer H. Abrams, *The Mirror and the Lamp: Romantic Theory and The Critical Tradition* (New York: Norton, 1958), esp. pp. 156–183; G. N. G. Orsini, "The Organic Concept in Aesthetics," *Comparative Literature* 31: 1969, pp. 1–30. Donald Drew Egbert provides an excellent account of use of the organic analogy in architecture in "The Idea of Organic Expression in American Architecture" in Stow Persons, ed. *Evolutionary Thought in America* (New Haven: Yale, 1950), pp. 336–396. Egbert relates the idea to functionalism and more significantly but less convincingly to three general philosophies of architecture: humanism, theocentrism and naturalism. The distinctions between the three are not quite as clear as Egbert suggests. However, his elaboration on the organic idea as an expression of naturalism is perhaps the most complete in tracing the sources and influences in the idea.

8. Leone Battista Alberti, *Ten Books on Architecture* (London: Alec Tiranti, 1955).

9. Bionics has been defined as the science of systems and devices which function in a manner characteristic of or resembling living systems. For an example of the thinking in this area, see J. I. Kremyanskiy, "Certain Peculiarities of Organisms as a 'System' from the Point of View of Physics, Cybernetics and Biology," and G. Sommerhoff, "The Abstract Characteristics of Living Systems"; F. E. Emery, ed., *Systems Thinking* (Baltimore: Penguin Books, 1969), pp. 125–202.

10. Richard P. Adams, "Architecture and the Romantic Tradition: Coleridge to Wright," *American Quarterly* 9(1): Spring 1957, pp. 46–62, 47.

11. Becker and Barnes, *Social Thought,* II, p. 531.

12. Morton and Lucia White, *The Intellectual Versus the City* (Cambridge: Harvard University Press, 1962), p. 235; Adams, "Architecture and the Romantic Tradition: Coleridge to Wright," p. 47; Sherman Paul, *Louis Sullivan: An Architect*

*in American Thought* (Englewood Cliffs, N.J.: Prentice-Hall, 1962), passim. The use and abuse of the organic analogy during the development of these figures has a fascinating history. See: M. H. Abrams, *The Mirror and the Lamp;* S. T. Coleridge, "On the Principles of Genial Criticism" (1814), in J. Shawcross, ed., *Biographia Literaria.* 2 vols. (London, 1907), vol. II, pp. 238–239; Richard H. Fogle, "Organic Form and American Criticism: 1840–1870," in Floyd Stovall, ed., *Development of American Literary Criticism* (Chapel Hill: University of North Carolina Press, 1955); G. MacKenzie, *Organic Unity in Coleridge* (Berkeley: University of California Press, 1939); J. Benziger, "Organic Unity, Leibniz to Coleridge," *Publications of the Modern Language Association of America,* 66 (1951), pp. 24–48; C. Lord, "Organic Unity Reconsidered," *Journal of Aesthetics and Art Criticism,* 52 (1964), pp. 263–268; Daniel Stempel, "Coleridge and Organic Form: The English Tradition," *Studies in Romanticism,* 6 (1967): pp. 89–97; Joseph Malof, "Meter as Organic Form," *Modern Language Quarterly,* 27 (1966): pp. 3–17; Jan Mikarovsky, "The Notion of Wholeness in the Theory of Art," *English Studies,* 5 (1968): pp. 173–183.

13. Charles Goodman and Wolf von Eckardt, *Life for Dead Spaces* (New York: Harcourt, Brace and World, 1963), p. 39.

14. Robert A. Nisbet, *The Quest for Community* (New York: Oxford University Press, 1953), p. 4.

15. White, *Intellectual Versus the City,* p. 162.

16. Walter Laquer's review of: G. L. Mosse, "The Crisis of German Ideology: Intellectual Origins of the Third Reich" and P. G. J. Pulzer, "The Rise of Political Anti-Semitism in Germany and Austria" in *The New York Review of Books,* 14 January 1965. See also Barbara Lane, *Architecture and Politics in Germany 1918–45* (Cambridge: Harvard University Press, 1968) and Clair Otto, "Germany and Nazis," *Journal Society Architectural Historians,* March 1965.

17. Nisbet, *Quest for Community,* pp. 10, 26, 29.

18. Kingsley Davis, *Human Society* (New York: Macmillan, 1949), p. 342.

19. Gilbert Herbert, "The Organic Analogy in Town Planning," *Journal of the American Institute of Planners* 29 (3): August 1963, p. 198.

20. Ibid., p. 201; The various analogies perceived by Herbert correspond largely to use of the analogy in social thought. There is perhaps a closer tie with his cosmological depiction since the central thrust of the analogy's social use was to suggest order and meaning in society and the state: a search for ideal relationships among a society's "functional" parts. This is clearly and succinctly brought out, for example, in M. B. Foster's summary of Hegel's conception of "ideality" in the relationship of constituent parts within an organic body: the state (M. B. Foster, *The Political Philosophies of Plato and Hegel.* Oxford: Clarendon Press, 1968; reprint of the 1935 edition, p. 189).

> The "ideality" of its constituent parts is what differentiates an organism from an inorganic body, constituting them not parts, but members, or organs. An organ is different from a part in that its essence is relative to the whole which includes it, and that it can realize its own perfection only in performing its function in the whole. The domination of the whole over the action of its constituent members is the life of the organism; perfection of it is health and diminution of it is disease. Disease, according to Hegel, consists in the partial emancipation of one organ of the body from this control, so that it begins a life and development of its own. In doing this, the organ not only destroys the perfection of the whole, it loses its own perfection in precisely the same degree,

failing to that extent to be what it is its essence to be. In an organism there can be no conflict between the interest of the part and that of the whole, because the proper being of the part is to be an organ of the whole and its proper activity to perform a function within it.

21. Ibid., p. 203.

22. Ibid., p. 204.

23. Ibid., p. 205.

24. Bruno Zevi, *Towards an Organic Architecture* (London: Faber and Faber 1950), pp. 67–70.

25. Ibid., pp. 69–70.

26. Walter Curt Behrendt, *Modern Building: Its Nature, Problems, and Forms* (New York: Harcourt, Brace, and Co., 1937), p. 11.

27. Ibid., p. 8.

28. Ibid., p. 73.

29. Ibid., pp. 9–17.

30. John Burchard and Albert Bush-Brown, *The Architecture of America* (Boston: Little Brown, 1961), pp. 20–21.

31. Eliot F. Noyes, *Organic Design in Home Furnishings* (New York: Museum of Modern Art, 1941), cover.

32. Perhaps it was only a superficial association between the two; the architect finding simply ornamental suggestions in nature. Hugh Morrison says of Sullivan, for example:

*Gray's Botany* influenced his ornament more than any other single source. He had a dog-eared copy, showing extensive use in studying the morphology of plants and their curious and marvellous differentiations within species. He referred the book to students frequently. His sketch-book was full of drawings from this source: complex organic developments from single germinal ideas.

Hugh Morrison, *Louis Sullivan: Prophet of Modern Architecture* (Westport, Conn.: Greenwood Press, 1971; first published by W. W. Norton, 1935), p. 225.

33. Sherman Paul, *Louis Sullivan: An Architect in American Thought* (Englewood Cliffs, New Jersey: Prentice-Hall, 1962), p. 101; and Gilbert Herbert, "Form and Function: A Study of Frank Lloyd Wright's Theory of Organic Architecture," *South African Architectural Record,* 44(9): September 1959, pp. 23–39, 29.

34. Louis H. Sullivan, *Kindergarten Chats and Other Writings* (New York: George Wittenborn: 1947; revision of 1918 edition), pp. 119, 160.

35. Paul, *Louis Sullivan,* p. 130; Sullivan, *Kindergarten Chats,* pp. 46–47.

36. For further elaboration on this association, see the first chapter of Edward R. DeZurko, *Origins of Functionalist Theory* (New York: Columbia University Press, 1957).

37. Herbert, "Form and Function: A Study of Frank Lloyd Wright's Theory," pp. 31–32. In this, Zevi is a slavish follower of Wright.

38. Frederick Gutheim, ed., *Frank Lloyd Wright on Architecture* (New York: Duell, Sloan and Pearce, 1941), p. 185; it was the spatial expression of this functional integration that was of ultimate consequence; see p. 189.

39. Lewis Mumford, *The City in History* (New York: Harcourt, Brace and World, 1961), p. 174.

40. Ibid., p. 444.

41. Ibid., p. 302.

42. Ibid., p. 303. See also Camillo Sitte (as in note 3, chapter 1), Eliel Saarinen,

*The City: Its Growth, Its Decay, Its Future* (New York: Reinhold, 1943), pp. 41, 44, 68, 74–79, 123.

43. Mumford, *City in History*, p. 52. Also, Lewis Mumford, "Social Complexity and Urban Design," *Architectural Record*, February 1963, p. 119; see also p. 123.

44. Lewis Mumford, "The Beginnings of Urban Integration," *Architectural Record*, January 1963, p. 119; "The Disappearing City," *Architectural Record*, October 1963, pp. 123, 125. In Mumford's view, Ebenezer Howard was the only planner to appreciate fully these perceptions which, of course, led him to an organic (cellular) form for the city. See: Mumford, "The Disappearing City," p. 123; "The Beginnings of Urban Integration," pp. 119, 126; "Social Complexity and Urban Design," p. 126.

45. Introduction by Gropius to Gilbert Herbert, *The Synthetic Vision of Walter Gropius* (Johannesburg: Witwaterstrand University Press, 1959), p. viii.

46. Herbert, *Synthetic Vision of Walter Gropius,* p. 23; see also p. 25.

47. Christopher Rand, "The Ekistic World," *New Yorker,* 11 May 1962, p. 26.

48. Diana Rowntree, "Second Delos Symposium," *Architectural Design* 34, September 1964, p. 425.

49. Another illustration of this is the work of Maurice E. H. Rotival; see particularly: "An Experiment in Organic Planning," *USA Tomorrow* 1 (2), January 1959, p. 19. See also Maurice E. H. Rotival, "Une Nouvelle Methode de Planification," *Urbanisme* No. 76: 1963, pp. 8–21; and for an application of the concept, Maurice E. H. Rotival, *Action for the Quinebaug Valley* (Hartford, Conn.: Redevelopment Commission, n.d.).

50. Eldon L. Modisette, "The Legitimization of Modern American Architecture," *Journal of Aesthetics and Art Criticism* 20 (3): Spring 1962, pp. 251–261.

51. Carl L. Becker, *The Heavenly City of the Eighteenth Century Philosophers* (New Haven: Yale University Press, 1959).

52. See De Zurko, *Origins of Functionalist Theory,* for a basic account of the functionalist vision and Peter Blake, "Paolo Soleri's Visionary City," *Architectural Record,* February 1961, pp. 111–118 for an example of this.

53. Richard Neutra, "Non-Visual Aspects of City Planning," in *Annual of Architecture, Structure and Town Planning* (Calcutta: Publishing Corp. of India, 1960), p. A–17; see also his statements in "Report on Harvard Urban Design Conference," *Progressive Architecture,* August 1956, esp. p. 98.

54. Bruno Zevi, *Towards an Organic Architecture,* p. 73. It must be noted also that the use of nature as a guide to the planning of urban systems is not restricted to a historical romanticism on the part of architects. For example, Athelston Spilhaus cited nature as model in the planning of his experimental city in an address to the American Association for the Advancement of Science, Dallas, Texas, 29 December 1968.

55. See, for example, Philip Boardman, *Patrick Geddes* (Chapel Hill: University of North Carolina Press, 1944), p. 80.

56. Walter Buckley, *Sociology and Modern Systems Theory* (Englewood Cliffs, N.J.: Prentice-Hall, 1967), pp. vii, 1, 7, 11–17.

57. Amos H. Hawley, *Human Ecology* (New York: Ronald Press, 1950), p. 51.

58. Alfred L. Kroeber, "The Superorganic" in Lewis A. Coser and Bernard Rosenberg, *Sociological Theory* (New York: Macmillan, 1957), p. 32.

59. Rudolf Heberle, In the Introduction to Ferdinand Tonnies, *Community and Society* (New York: Harper Torchbooks, 1957), p. ix.

60. Ludwig von Bertalanffy, *Problems of Life* (New York: Wiley, 1952), pp. 11, 273.

61. W. Ross Ashby, *An Introduction to Cybernetics* (New York: Wiley, 1963), pp. 1–6.

62. Norbert Wiener, "Cybernetics," in Alfred G. Smith, ed., *Communication and Culture* (New York: Holt, Rinehart and Winston, 1966), p. 27.

63. Ralph E. Gibson, "The Recognition of Systems Engineering," in Charles D. Flagle, ed., *Operations Research and Systems Engineering* (Baltimore: Johns Hopkins Press, 1960), p. 58.

64. Ibid., p. 59.

65. Howard Becker and Harry Elmer Barnes, *Social Thought from Lore to Science* (3 vols.; New York: Dover, 1961), vol. 3, pp. 680–681.

66. Frederick Gutheim, *Frank Lloyd Wright on Architecture,* pp. 197–198.

67. John Wesley Bookwalter, *Rural Versus Urban: Their Conflict and Its Causes* (New York: Knickerbocker Press, 1910), pp. 4–6.

68. Paul Jacques Grillo, *What is Design?* (Chicago: Paul Theobald, 1960), passim.

69. Gunther Nitschke, "The Metabolists of Japan," *Architectural Design* 34 (10): October 1964, pp. 509–524.

70. Fumihiko Maki, "Dojima Redevelopment Plan," *Japan Architect* 38: pp. 75–80, May 1963, p. 73; also, one may wish to compare these "metabolic" expressions of the idea in planning to the systems-engineering approach and, more specifically, to the idea of nature as "designer" as seen in bionics. An example of the latter is contained in Howard C. Howland, "Structural, Hydraulic and 'Economic' Aspects of Leaf Ventation and Shape," in Eugene E. Bernard and Morley R. Kare, eds., *Biological Prototypes and Synthetic Systems* (New York: Plenum Press, 1962).

71. The ideological associations of the organic idea have been eschewed in this essay for the sake of brevity. They form no coherent pattern. Charles Elwood, for example (*History of Social Philosophy,* pp. 463–477), in exploring the idea's use by Spencer, von Lilienfeld, and Schaeffle finds associations with extreme conservatism as well as moderate progressivism. To some it is simply a fascist idea, e.g., Ross S. J. Hoffman, *The Organic State: An Historical View of Contemporary Politics* (New York: Sheed and Ward, 1939), pp. 5, 13, 26, 37; while others would add a Marxist tag, e.g., Nicholas S. Timasheff, *The Sociology of Luigi Sturzo* (Baltimore: Helicon Press, 1962), p. 168. The minimal consistency that does exist in the idea's ideological associations stems from an abstract and nebulous holism with the strong interdependence of parts necessary to the social whole—implying some form of collectivism. This thinking is usually contrasted with a "social contract" or "natural rights" view of society in which the individual is seen as a relatively independent agent. See: Coker, *Organismic Theories of the State,* pp. 9–10. R. W. Gerard, through a biological lens, sees the force of all social evolution directed towards a greater integration of all the functional "organs" or subsystems of society. This will require an increasing subordination of individuals within society for the benefit of society's overall or "system" goals, its ends as a collectivity or organism. This he sees accomplished through a willing subordination of individual goals rather than through any coercive suppression: R. W. Gerard, "Higher Levels of Integration," in Redfield, ed., *Levels of Integration,* p. 82. Similar biological connotations of the idea suggest socialism or communism, e.g., Joseph Needham, *Order and Life* (Cambridge: The University Press, 1936); Henry

Maudsley, *Organic to Human: Psychological to Sociological* (London: Macmillan, 1916), pp. 189–225. Then, of course, one can get into the endless debate as to whether Plato's organic model of the state was communist, fascist, or democratic, e.g., Elwood, *History of Social Philosophy,* pp. 33–34; and Barker, *Greek Political Theory,* pp. 272–275. The ideological connotation becomes no clearer when organicism sheds its biological garb and becomes modern systems theory. Martin Kuenzlen, *Playing Urban Games: The Systems Approach to Planning* (Boston: The i Press, 1972), finds it a manipulative capitalist device while V. A. Lektorsky and V. N. Sadovsky, "On Principles of Systems Research," *General Systems Yearbook* 5: 1960, pp. 171–179, see it as Marxist. Cf. Whitaker, "Nature and Value of Functionalism," p. 132 for this association in a "functionalist" format.

72. Morton and Lucia White, *The Intellectual Versus the City* (Cambridge: Harvard University Press, 1962), pp. 235–236.

73. Lowdon Wingo, Jr., ed., *Cities and Space* (Baltimore: Johns Hopkins University Press, 1963), pp. 4–7.

74. Paul Goodman, "Columbia's Unorthodox Seminars," *Harper's Magazine,* January 1964, pp. 74–75.


## 6.   As a Guide to Assessment and Choice


1. William L. C. Wheaton and Margaret F. Wheaton, "Identifying The Public Interest: Values and Goals," in Ernest Erber, ed., *Urban Planning in Transition* (New York: Grossman, 1970), pp. 152–164; Paul Davidoff, "Advocacy and Pluralism in Planning," in *Journal of The American Institute of Planners* 31(4), November 1965.

2. For a brief history see: Virginia Held, "PPBS Comes to Washington," and Allen Schick, "The Road to PPB: The Stages of Budget Reform," in James W. Davis, Jr., ed., *Politics, Programs and Budgets* (Englewood Cliffs, N.J.: Prentice-Hall, 1969), pp. 138–149, 210–229.

3. For an excellent presentation of the central issues in making these assessments in the public sector, see: Roland N. McKean, *Public Spending* (New York: McGraw-Hill, 1968) and Alan Williams, "The Optimal Provision of Public Goods in a System of Local Government," in *Journal of Political Economy,* February 1966, pp. 18–33.

4. For some interesting recent techniques not discussed here see: Joseph S. DeSalvo, "A Methodology for Evaluating Housing Programs," in Arnold E. Harberger et al., eds., *Benefit-Cost Analysis 1971,* pp. 348–360 (Chicago: Aldine-Atherton, 1972); James R. Miller, *Professional Decision-Making: A Procedure for Evaluating Complex Alternatives* (New York: Praeger, 1970); Burton V. Dean and Samuel J. Mantel, Jr., "A Model for Evaluating Costs of Implementing Community Projects," in Mark Alfandary-Alexander, ed., *Analysis For Planning-Programming-Budgeting* (Washington: Washington Operations Research Council, 1968), pp. 27–47; Hans Dehlinger and Jean-Pierre Protzen, *Debate and Argumentation in Planning* (Berkeley: University of California, Institute of Urban and Regional Development, Working Paper No. 178, 1972); Derek Medford, *Environmental Harassment or Technology Assessment* (New York: Elsevier, 1973); Nestor E. Terleckyj, "Estimating Possibilities for Improvement in The Quality of Life," in *Looking Ahead* (Washington: National Planning Association) 20(10): January 1973; See the assessment

of these techniques in: A. Charnes, et al., "Measuring, Monitoring and Modeling Quality of Life," in *Management Science*, 19(10), June 1973, pp. 1172–1188. See also the proposal: "Decision Making and The Cost-effectiveness Criterion," in Karl Seiler, *Introduction to Systems Cost-Effectiveness* (New York: Wiley-Interscience, 1969), pp. 96–101.

5. A standard text in the field is Eugene L. Grant and W. Grant Ireson, *Principles of Engineering Economy*, 5th edition (New York: Ronald Press, 1970). For illustrations of architectural and urban planning use see: Rodolfo J. Aguilar, *Systems Analysis and Design in Engineering, Architecture, Construction and Planning* (Englewood Cliffs, N.J.: Prentice-Hall, 1973), pp. 11–44, and Robley Winfrey, *Economic Analysis for Highways* (Scranton, Pa.: International Textbook Co., 1969).

6. Aguilar, *Systems Analysis and Design*, p. 15.

7. Also, the three provide sufficient background to begin an introduction to cost-benefit analytic techniques in architecture and urban planning. For the prospective rate of return method, see: Grant and Ireson, *Principles of Engineering Economy*, pp. 109–134.

8. For the appropriate transformations, see ibid., p. 594.

9. This is a very contentious issue in the analysis. For an overview, see: William J. Baumol, "On The Discount Rate for Public Projects," in Robert H. Haveman and Julius Margolis, eds., *Public Expenditures and Policy Analysis* (Chicago: Markham, 1970), pp. 273–289; Jacob A. Stockfisch, "The Interest Rate Applicable to Government Investment Projects," in Harley H. Hinrichs and Graeme M. Taylor, *Program Budgeting and Benefit-Cost Analysis* (Pacific Palisades, Calif.: Goodyear, 1969), pp. 187–201.

See also the various papers in Part 3 of: "The Social Time Preference Rate and The Social Opportunity Cost of Capital," in Richard Layard, ed., *Cost-Benefit Analysis* (Baltimore: Penguin Books, 1972), pp. 243–332.

10. A. M. Wellington, *The Economic Theory of Railway Location* (New York: Wiley, 1887), cited in Grant and Ireson, *Principles of Engineering Economy*, p. 3.

11. To quote:

It is hereby recognized that destructive floods upon the rivers of the United States, upsetting orderly processes and causing loss of life and property, including the erosion of lands, and impairing and obstructing navigation, highways, railroads, and other channels of commerce between the States, constitute a menace to national welfare; that is the sense of Congress that flood control on navigable waters or their tributaries is a proper activity of the Federal Government in cooperation with States, their political subdivisions, and localities thereof; that investigations and improvements of rivers and other waterways, including watersheds thereof, for flood-control purposes are in the interest of the general welfare; that the Federal Government should improve or participate in the improvement of navigable waters or their tributaries, including watersheds thereof, for flood-control purposes if the benefits to whomsoever they may accrue are in excess of the estimated costs, and if the lives and social security of people are otherwise adversely affected.

Flood Control Act of 22 June 1936, United States Code, 1940 ed. (Washington: U.S. Government Printing Office, 1940), p. 2964, quoted in Grant and Ireson, *Principles of Engineering Economy*, p. 135

12. Samuel B. Chase, Jr., ed., *Problems in Public Expenditure Analysis* (Washington: The Brookings Institution, 1968), pp. 2–3.

13. See note 34 in Chapter 4 for a critique of the analysis.

14. H. G. Walsh and Alan Williams, *Current Issues in Cost-Benefit Analysis* (London: H.M.S.O., Civil Service Department, CAS Occasional Paper No. 11, 1969), p. 6.

15. Michael B. Tietz, "Cost Effectiveness: A Systems Approach to Analysis of Urban Services," *Journal of the American Institute of Planners,* 34(5), September 1968, p. 307 (emphasis added).

16. Nathaniel Lichfield, "Cost-Benefit Analysis in Plan Evaluation," *Town Planning Review* 35, July 1964, pp. 160–169. For a fuller explanation of Lichfield's concept, see: his "Cost-Benefit Analysis in City Planning," *Journal of The American Institute of Planners,* 26, 1960, pp. 273–279; "Evaluation Methodology of Urban and Regional Plans: A Review," *Regional Studies* 4, 1970, pp. 151–165; "Cost-Benefit Analysis in Planning: A Critique of The Roskill Commission," in *Regional Studies,* Vol. 5, 1971, pp. 157–183.

17. A. R. Prest and Ralph Turvey, "Cost-Benefit Analysis: A Survey," *The Economic Journal* 75: December 1965, pp. 718–719.

18. To quote:

How valid is the balance sheet for its proposed purpose? How useful is it for the evaluation of alternative courses of action? The costs and benefits that are enumerated refer to different objectives and are not all relevant for a single objective. For instance, in the analysis of the costs and benefits of the retention of the San Francisco Mint, the historic value of the Old Mint to San Francisco and the nation, the economic contribution to the San Francisco economy of visitors to the Mint, and the value of the office space to the Federal Government are not relevant for any single objective, unless that objective is the catch-all and consequently meaningless one of "enhancing community welfare." Thus a major criticism of the "development balance sheet" is that it does not appear to recognize that benefits and costs have only instrumental value. Benefits and costs have meaning only in relation to a well-defined objective. A criterion of maximizing net benefits in the abstract is therefore meaningless. Whereas benefits can be computed referring to different planning objectives, the benefits and costs are not necessarily additive or comparable. It is meaningful to add or compare benefits and costs only if they refer to a common objective. Furthermore, since benefits and costs can legitimately be compared only in terms of an objective, if the objective is of little or no value both for an entire community and for any sections within it, then the benefits and costs referring to the objective are irrelevant for the community in question. For instance, if a community as a whole and all interests within it set no value on the retention of historic buildings, it is not legitimate for an analyst to consider the elimination of a building of historic value as a cost even though he personally believes it to be so. Morris Hill, "A Goals-Achievement Matrix for Evaluating Alternative Plans," *Journal of the American Institute of Planners,* 34(1): January 1968, pp. 19–29, 21.

19. See note 3, Chapter 10.

20. Prest and Turvey, "Cost Benefit Analysis: A Survey," pp. 730–731.

21. Despite the many recent elaborations of analytic techniques, some fundamental problems still exist. One is the lack of a holistic vision on the part of persons responsible for program decisions. As Daniel P. Moynihan has noted (Daniel P. Moynihan, ed., *Toward a National Urban Policy* [New York: Basic Books, 1970], pp. 8–9), we must

develop a much heightened sensitivity to . . . "hidden" urban policies. There is hardly a department or agency of the national government whose programs do not in some way have important consequences for the life of cities, and those who live in them. Frequently—one is tempted to say normally!—the political appointees and career executives concerned do not see themselves as involved with, much less responsible for the urban consequences of their programs and policies. They are, to their minds, simply building highways, guaranteeing mortgages, advancing agriculture, or whatever. No one has made clear to them that they are simultaneously redistributing employment opportunities, segregating neighborhoods, or desegregating them, depopulating the countryside and filling up the slums, and so forth: all of these things are second and third order consequences of nominally unrelated programs. Already this institutional naïveté has become cause for suspicion; in the future it simply must not be tolerated. Indeed, in the future, a primary mark of competence in a federal official should be the ability to see the interconnections between programs immediately at hand, and the urban problems that pervade the larger society.

### 7. At the City Scale: Modeling Large Systems

1. Arturo Rosenblueth and Norbert Wiener, "The Role of Models in Science," *Philosophy of Science* 12(4): October 1945, pp. 316–321.

2. For an excellent introduction to those techniques in urban planning, see: Philip M. Morse, ed., *Operations Research for Public Systems* (Cambridge: MIT Press, 1967); Rodolfo J. Aguilar, *Systems Analysis and Design* (see Chapter 6, note 5); and the works cited in note 15 of Chapter 2.

3. For a summary of the functions of models and the contexts of their use, see: "Models: General Framework," in Benjamin Reif, *Models in Urban and Regional Planning* (New York: Intertext, 1973), pp. 49–60; Maurice D. Kilbridge, et al., "A Conceptual Framework for Urban Planning Models," in *Management Science* 15(6) February 1969; Maurice D. Kilbridge and S. Carabateas, "Urban Planning Models: A Classification System," in *Ekistics* 24(145): December 1967, pp. 480–485; A. G. Wilson, *Models in Urban Planning: A Synoptic Review of the Literature* (London: Center for Environmental Studies, 1968; Working Paper CES–WP–3).

4. See, for example: Morton Schneider, "Access and Land Development," and George C. Hemmens, "Survey of Planning Agency Experience with Urban Development Models, Data Processing, and Computers," in George C. Hemmens, ed., *Urban Development Models* (Washington: Highway Research Board, 1968; Special Report No. 97), pp. 164–177, 219–230. For an assessment of the "real" world utility of the models as expressed by their operational use, see: Janet Rothenberg Pack, *The Use of Urban Models: Report of a Survey of Planning Organizations* (Philadelphia: University of Pennsylvania/Fels Center of Government, 1973; Fels Discussion Paper No. 42); Janet Rothenberg Pack, et al., *The Use of Urban Models in Urban Policy Making* (Philadelphia: University of Pennsylvania/Fels Center of Government, 1974, 2 volumes).

5. George B. Dartzig and Thomas L. Saaty, *Compact City: A Plan for a Liveable Urban Environment* (San Francisco: W. H. Freeman, 1973).

6. Jay W. Forrester, *Urban Dynamics* (Cambridge: MIT Press, 1969).

7. Forrester uses many "best guesses" to define specific and critical relationships among the significant variables. Forrester, ibid., p. 114. His justification is that:

much of the behavior of systems rests on relationships and interactions that are believed, and probably correctly so, to be important but that for a long time will evade quantitative measure. Unless we take our best estimates of these relationships and include them in a system model, we are in effect saying that they make no difference and can be omitted. It is far more serious to omit a relationship that is believed to be important than to include it at a low level of accuracy that fits within the plausible range of uncertainty. In this particular aspect the kind of modeling discussed here follows the philosophy of the manager or political leader more than that of the scientist. If one believes a relationship to be important, he acts accordingly and makes the best use he can of the information available. He is willing to let his reputation rest on his keenness of perception and interpretation.

For some persuasive cautions on this point see: William Alonso, "The Quality of Data and The Choice and Design of Predictive Models," in George C. Hemmens, ed., *Urban Development Models,* pp. 178–192.

8. The most comprehensive critique is that provided by Garn and Wilson, in a footnote which suggests its thrust (Harvey A. Garn and Robert H. Wilson, *A Critical Look at Urban Dynamics: The Forrester Model and Public Policy* [Washington: The Urban Institute, 1970], footnote 10, p. 21).

In *Urban Dynamics,* Professor Forrester cites a total of six references—none of which is a part of the rapidly growing literature in the field of urban studies. In the Preface, Professor Forrester justifies his approach because he expected "the most valuable source of information to be, not documents, but people with practical experience in urban affairs" (p. ix). Without wishing to disparage their contribution, it almost goes without saying that the gap between experience gained in practice and the knowledge that research produces is often alarmingly large and in favor of research. One cannot escape the feeling that reference to the research done in this area would have led to significant changes in the specification of many of the functional relationships in *Urban Dynamics.*

See also: Gregory K. Ingram and John F. Kain, *Two Views of Urban Dynamics* (Cambridge: Harvard University, Program on Regional and Urban Economics, 1970; Discussion Paper No. 71). The Ingram paper was first published in the AIP Journal, May 1970; Kain's in *Fortune,* November 1969; Leo P. Kadanoff, "An Examination of Forrester's Urban Dynamics," in *Simulation,* June 1971, pp. 261–268; Peter Passell, Marc Roberts, and Leonard Ross, "Review of Urban Dynamics," in *New York Times,* 2 April 1972; a highly politicized critique appears in Martin Kuenzlen, *Playing Urban Games: The Systems Approach to Planning* (George Braziller, 1972), pp. 74–79, 87.

9. To place Forrester's model in the context of other large-scale modeling efforts and to appreciate the difficulties of all enterprises of that genre, see the cogent critique by Douglass B. Lee, Jr., "Requiem for Large-Scale Models," in *Journal of The American Institute of Planners:* May 1973, pp. 163–178.

## 8. At the Architectural Scale: Systems Building and Building Systems

1. Ian Donald Terner and John F. C. Turner, *Industrialized Housing* (Washington: U.S. Agency for International Development and Department of Housing and Urban Development, 1972), pp. I-5, I-6.

2. For a brief history of systems building, see: Thomas Schmid and Carlo Testa,

*Systems Building: An International Survey of Methods* (New York: Praeger, 1969), pp. 28–33.

3. Of his system, Le Corbusier said:

We designed a structural system, a frame, completely independent of the functions of the plan of the house: this frame simply supports the flooring and the staircase. It is made of standard elements which can be fitted together, thus permitting a great diversity in the grouping of the houses. . . . At the request of the town-planner or customer, such frames, oriented and grouped, can be delivered by a manufacturer anywhere in the country. . . .

It then remained to fit up a home inside these frames. . . . We conceived the idea of a firm, an affiliate of the first, which would sell all the elements required to equip the house, everything, that is, which can be manufactured, mass-produced in standard sizes, and meet the various needs of a rational installation: windows, doors, standard casings serving as cupboards and forming part of the dividing walls. . . . Since the Dom-ino framework bore the loads, the walls and partitions could be made of any material.

Maurice Besset, *Who Was Le Corbusier?* trans. Robin Kemball (Geneva: Editions d'Art Albert Skira, 1968), p. 69.

4. Le Corbusier and Pierre Jeanneret, *Oeuvre Complete 1910–1929* (Zurich: Les Editions D'Architecture, 1964), pp. 23–26, 45–47.

5. Burnham Kelly, *The Prefabrication of Houses* (Cambridge and New York: MIT Press and John Wiley and Sons, 1951), pp. 26–28.

6. Schmid and Testa, *Systems Building: An International Survey,* pp. 34–36.

7. One of the first successful introductions of a large-scale European housing system in the United States was that by Module Communities, Inc. of The Tracoba System widely used in France. Construction on a twenty story structure in Yonkers, New York began in 1970. The project was directed by Professor Harold K. Bell of Columbia University.

8. "Industrialized Housing: What is it . . . really—and where is it going?", *House and Home,* 44(5): November 1973, pp. 65–66. The *House and Home* definition of industrialized housing is also interesting:

*First, it's built almost entirely off the site.* Walls, floor, roof, kitchen, baths and all exterior and interior finishing are built, assembled and finished in a factory, and then trucked to the site as three-dimensional modules, or sections. *Second, it's built by relatively unskilled labor, which means the assembly line.* Usually working in the $3-to-$4 an hour range, this labor performs repetitive functions under controlled factory conditions. *And, third, it's standardized because that's the only way it can be put on an assembly line.* Basic dimensions and configurations remain constant because variations, other than very minor ones, slow down the production line.

9. "Building Systems Information Clearinghouse," *Building Systems Planning Manual* (Menlo Park, Calif.: Educational Facilities Laboratories, Inc., 1971), p. 3.

10. Ibid, p. 2; and C. W. Griffin, Jr., *Systems: An Approach to School Construction* (New York: Educational Facilities Laboratories, Inc., 1971).

11. "Sweets Refines Logic of Product Search," *Architectural Record,* 154(2), August 1973, pp. 65–66; Miriam S. Eldar, "Product Information: Sweet's Guidelines Structure," *Industrialization Forum,* 4(4), 1973, pp. 33–46.

12. Laurence S. Cutler, "The Industrialized Evolution" in Albert G. H. Dietz and Laurence S. Cutler, *Industrialized Building Systems for Housing* (Cambridge: MIT Press, 1971), p. 101; John G. Brandenburg, *The Industrialization of Housing: Im-*

*plications for New Town Development* (Chapel Hill: Center for Urban and Regional Studies, University of North Carolina, 1970), p. 4. See also: E. Jay Howenstine, *Productivity Trends in The Construction Industry: A Comparative International Review* (Washington: Department of Housing and Urban Development, 1973; reprinted from: *Measuring Productivity in The Construction Industry,* National Commission on Productivity, 1973).

13. Robert E. Platts, "System Housing: The Shelter Industry Shapes Up," in Dietz and Cutler, *Industrialized Building Systems,* p. 121.

14. "What's Holding Back the Real Breakthrough in Modular Housing?" *House and Home,* 40(4): October 1971, p. 96.

15. A summary of the case appears in the *New York Times,* 18 April 1967, p. 1 along with the reasoning behind the Supreme Court's decision in favor of the unions.

16. Gunnar Myrdal, *Journal of Housing,* No. 8, September 1967, as quoted in *Industrialized Housing* (Committee Print, Sub-Committee on Urban Affairs, Joint Economic Committee, 91st Congress; Washington: U.S. Government Printing Office, April 1969), p. 167. Myrdal said:

> First, of all, continuity of operation is required if the building industry is to amortize any great increase in capital investment. The industry cannot be expected to take the risk of employing highly capital-intensive methods of production as long as governments keep using housing construction as a regulator of the national economy. This point has been made before but is worth repeating. There are few sectors of the economy less suited for economic balancing than housing production—unless, happily, if it is lifted to industrialized building. If construction activity has to be used for economic balancing purposes—which may occasionally be necessary, though less so with wiser economic policies applied by the governments—the projects for periodic retrenchment should be searched for in other sectors than housing: in urban renewal projects, public works, and other demands of a once-for-all type. The most important incentive towards industrialized building would be a guarantee on the part of the government that mass construction of residential buildings will not be interfered with but everything done in order to make possible a steady, rising level of housing construction.

See also the accompanying document on the July 1969 hearings by the same Committee, also entitled *Industrialized Housing,* for a useful summary of the state of the art in Industrialized Housing.

17. Alberto F. Trevino, Jr., "Some Insights into Systems Building," in *Urban Land,* July–August 1970, p. 4.


## 9. In Systematizing the Planning and Design Process

1. The range of authorship in two basic works in this rapidly developing area testifies to our point: Gary T. Moore, ed., *Emerging Methods in Environmental Design and Planning* (Cambridge: MIT Press, 1970); J. Christopher Jones, *Design Methods: Seeds of Human Futures* (New York: Wiley-Interscience, 1970). This is also sustained by perusal of any issue of the *DMG-DRS Journal: Design Research and Methods.*

2. See note 7 Chapter 1. Is it any easier to specify what is wanted in the physical environment? See: William Michelson, "Most People Won't Want What

Architects Want," *Transaction,* vol. 5 (July–August 1968), pp. 37–43; Clare Cooper and Phyllis Hackett, *Analysis of the Design Process at Two Moderate-Income Housing Developments* (Berkeley, Calif.: Center for Planning and Development Research, University of California, 1968).

3. The "California" studies were early applications of the analytic process: Lockheed Missiles and Space Company, *Statewide Information System Study,* 3 vols. (Sunnyvale, Calif.: Lockheed, 1965); Aerojet-General Corporation, *Waste Management Study* (Azusa, Calif.: Aerojet-General, 1965); North American Aviation, Inc., *Integrated Transportation Study,* 5 vols. (Los Angeles: North American, 1965); Space-General Corporation, *Prevention and Control of Crime and Delinquency* (El Monte, Calif.: Space-General, 1965). See the critique of same in: Ida R. Hoos, *Systems Analysis in Social Policy* (London: Institute of Economic Affairs, 1969), pp. 38–51.

4. For an assessment of the "state of the art" now, see: Nigel Cross, "The Day The Music Died," in *DMG-DRS Journal* 6(4): October–December 1972, p. 185, and Eric Dluhosch, "What if . . . Some Thoughts on Methodology, Technology and The Scientific Approach—Viewed Through The Looking Glass of Berkeley," in *DMG-DRS Journal* 7(3): July–September 1973, pp. 219–228.

5. The context of such efforts and their central concepts are established in: Robert Sommer, *Personal Space: The Behavioral Basis of Design* (Englewood Cliffs, N.J.: Prentice-Hall, 1969); Harold M. Proshansky, William H. Ittelson, and Leanne G. Rivlin, "The Influence of the Physical Environment on Behavior: Some Basic Assumptions," in Harold M. Proshansky, William H. Ittelson, and Leanne G. Rivlin, eds., *Environmental Psychology: Man and His Physical Setting* (New York: Holt, 1970); Constance Perin, "Concepts and Methods for Studying Environments in Use," in William J. Mitchell, ed., *Environmental Design: Research and Practice: Proceedings of the EDRA3/AR8 Conference* (Los Angeles: University of California, 1972), 13.6.1–13.6.10; Kenneth H. Craik, "The Comprehension of the Everyday Physical Environment," *Journal of the American Institute of Planners,* vol. 34 (January 1968), pp. 29–37.

An excellent example at the architectural scale of how "needs" are determined and the physical artifact designed is: Alexander Kira, *The Bathroom: Criteria for Design* (Ithaca, N.Y.: Center for Housing and Environmental Studies, Cornell University, 1966).

6. Perceptions of "User Requirements" in housing will reflect significant cultural, social and economic variations in addition to the essentially psychological. See: Stanley Milgram, "The Experience of Living in Cities," *Science,* vol. 167 (March 13, 1970), 1461–1468; Amos Rapoport, *House Form and Culture* (Englewood Cliffs, N.J.: Prentice-Hall, 1969); W. L. Yancey, "Architecture, Interaction and Social Control: the case of a large scale public housing project," *Environment and Behavior,* vol. 3, no. 1 (1971), pp. 3–21; R. Maurer and J. C. Baxter, "Images of the neighborhood and city among Black-, Anglo-, and Mexican-American children," *Environment and Behavior,* vol. 4 (December 1972), pp. 351–388.

7. There are some outstanding exceptions. Of the broader, more comprehensive studies, one of the best is: J. K. Friend and W. N. Jessop, *Local Government and Strategic Choice* (Beverly Hills: Sage, 1969).

8. CRAFT is an acronym for "Computerized Relative Allocation of Facilities Techniques" (IBM SHARE program #3391). For an introduction to the program as well as suggestions for its improvement see: I. Paul Lew and Peter H. Brown, "Evaluation and Modification of CRAFT for an Architectural Methodology," in Gary T. Moore, *Emerging Methods,* pp. 155–161.

9. The family of models I have found most promising in their potential as a theoretical foundation for physical planning and design are those concerned with "Activity Systems." The basic idea here is that human activity is the generic functional determinant of form. A description of various activities leads ultimately to their probable physical consequences, and tracing the causal relationships provides the basis for a design methodology. As such, they might be viewed as a subset of a "user needs" model. In any event, they are interesting and, at the architectural scale best expressed in *Activity Date Method* (London: Her Majesty's Stationery Office, 1966); *Planning a Major Building Programme,* Ministry of Public Buildings and Works (London: Her Majesty's Stationery Office, 1966).

The urban-planning use of the idea is obviously more complex and consequently not yet fully developed. See: F. Stuart Chapin, Jr. and H. C. Hightower, "Household Activity Patterns and Land Use," *Journal of American Institute of Planners* 31 (3): August 1965, pp. 222–231; F. Stuart Chapin, Jr., *Urban Land Use Planning* (Urbana: University of Illinois Press, 1965), revised edition; F. Stuart Chapin, Jr., "Activity Systems and Urban Structure: A Working Schema," *Journal of American Institute of Planners* 34 (1): January, 1968, pp. 11–18. F. Stuart Chapin, Jr., "Activity Systems as a Source of Inputs for Land Use Models," in George C. Hemmens, *Analysis and Simulation of Urban Activity Patterns:* paper prepared for ACM October, 1968.

## 10. A General Assessment of the Systems Approach in Architecture and Urban Planning

1. Robert H. Haveman, Introduction, in Robert H. Haveman and Julius Margolis, eds., *Public Expenditure and Policy Analysis* (Chicago: Markham, 1970), p. 7.

2. Paul Diesing, "Noneconomic Decision-Making," in *Ethics,* 65(1): October 1954, pp. 18–35, 18.

3. It is the rigor employed in determining the economically "rational" that I feel leads the analyst to a spurious sense of having solved a social problem. The often elegant techniques do solve problems in marginal utility decisions (vide Diesing above) but what of the perception of the components of that utility? Since the point is significant in the following assessment, it is useful to summarize it here by the juxtaposition of the views of two distinguished and very perceptive observers of rationality in society: Irving Kristol and Kenneth J. Arrow. Kristol says in "The Corporation: A Last Word," *Wall Street Journal,* 14 March 1974, that

a good economist has a mind like a razor, which is why he is so useful—and so dangerous. He is useful because he can make sense of a tangled and otherwise bewildering situation. He is dangerous because he can make only economic sense of it—and the world does not move by economic sense alone.

This is a disconcerting statement in view of Arrow's assertion in Kenneth J. Arrow, *Limits of Organization* (Philadelphia: University of Pennsylvania/Fels Center of Government, 1973); First Annual Fels Lectures on Public Policy Analysis, Lecture 1: "Rationality: Individual and Social", pp. 1–2, 1–3, that

an economist by training thinks of himself as the guardian of rationality, the ascriber of rationality to others, and the prescriber of rationality to the social world. . . . [and later] Rationality, after all, has to do with means and ends and their relation. It does not specify what the ends are. It only tries to make us aware of the congruence or dissonance between the two. So ultimately any value discussion must come to a rest temporarily on unanalyzed postulates.

There is an infinite regress as we try to justify one value judgment in terms of supposedly deeper ones.

Gunnar Myrdal, however, insists in *An American Dilemma,* vol. 2 (New York: Harper and Row, 1969), p. 1044, on the social scientist's claim to particular insights:

The scientist—even if his knowledge is conjectural in certain re pects—is in a position to assist in achieving a much wiser judgment than the one which is actually allowed to guide public policy. . . . Nor can we argue that "the facts speak for themselves" and leave it "to the politician and the citizen to draw the practical conclusions." We know even better than the politician and the ordinary citizen that the facts are much too complicated to speak an intelligible language by themselves. They must be organized for practical purposes, that is, under relevant value premises. And no one can do this more adequately than ourselves.

Since the issue is critical, the reader may wish to see the context established in Lloyd G. Reynolds, *The Three Worlds of Economics* (New Haven: Yale, 1971), pp. 5–8, 14–17, 303–329.

4. E. S. Quade, *Analysis for Military Decisions* (Chicago: University of Chicago Press, 1964), p. 153. Aaron Wildavsky, in citing Quade in "The Political Economy of Efficiency: Cost-Benefit Analysis, Systems Analysis and Program Budgeting," *Public Administration Review* 26(4): December 1966, p. 299, stresses the point: "It cannot be emphasized too strongly that a (if not the) distinguishing characteristic of systems analysis is that the objectives are either not known or are subject to change" (p. 299). For the implications of this condition to the planner, we can go all the way back to Seneca (4 B.C.–65 A.D.). In his "Epistles to Lucilius" (LXXI,3), he noted that: "Our plans miscarry because they have no aim. When a man does not know what harbor he is making for, no wind is the right wind."

5. Charles J. Hitch *On the Choice of Objectives in Systems Studies* (Santa Monica: RAND, 1960), p. 4 (emphasis added). See also his "An Appreciation of Systems Analysis," *Operations Research,* 3(4): November 1955, pp. 466–481.

6. Abram Bergson, *Economics of Soviet Planning* (New Haven: Yale University Press, 1964). Quoted in David Braybrooke and Charles E. Lindblom, *A Strategy of Decision: Policy Evaluation as a Social Process* (New York: The Free Press, 1963), p. 13.

7. Fred M. Frohock, *The Nature of Political Inquiry* (Homewood, Ill.: Dorsey Press, 1967), p. 157.

8. The mere fact, however, that we are willing to engage the problem has subjective implications. Paul Rogers writes on this point:

In any scientific endeavor—whether 'pure' or applied science—there is a prior subjective choice of the purpose or value which the scientific work is perceived as serving. This subjective value choice which brings the scientific endeavor into being must always be outside of that endeavor and can never become a part of the science involved in that endeavor. [Rogers accordingly asks:] Who will be controlled? Who will exercise control? What type of control will be exercised? Most important of all, toward what end or what purpose, or in the pursuit of what value will control be exercised?

C. R. Rogers and B. F. Skinner, "Some Issues Concerning The Control of Human Behavior," *Science,* 124: 1956, p. 1062, as quoted in C. Marshall Lowe, *Value Orientations in Counseling and Psycho-Therapy* (San Francisco: Chandler, 1969), p. 218.

9. Adm. Hyman Rickover in a hearing before the Committee on Foreign Relations on Defense Department Sponsored Foreign Affairs Research, 90th Congress, 2nd session, part 2 (May 2, 1968).

10. This is a somewhat selective criticism. More general, informed, and useful critiques of the analytic process can be found in: E. S. Quade, "Pitfalls and Limitations," in E. S. Quade and W. I. Boucher, eds., *Systems Analysis and Policy Planning* (New York: American Elsevier, 1969), pp. 345–363; G. D. Brewer, *Systems Analysis in the Urban Complex: Potential and Limitations* (Santa Monica: Rand Corporation, 1973; publication P–5141); Clay Thomas Whitehead, *Uses and Limitations of Systems Analysis* (Santa Monica: Rand Corporation, 1967; publication P–3683); John P. Mayberry "Broader Implications of PPB Systems," in Mark Alfandary-Alexander, ed., *Analysis for Planning-Programming-Budgeting* (Washington: Washington Operations Research Council, 1968), pp. 151–162.

To place these critiques in the context of the "rationality" of analysis see: "Utilitarian Calculation" and "Dewey's Instrumental Thinking," in Wayne A. R. Leys, *Ethics for Policy Decisions* (New York: Greenwood Press, 1968, first published by Prentice-Hall, 1952), pp. 13–32, 150–175.

11. C. E. M. Joad, *Guide to Modern Thought* (New York: Folcroft, 1973, repr. of 1933 ed.). It is difficult to assess quality even in the objective or "hard" systems areas. See, for example: R. J. Aumann and J. B. Kruskal, "Assigning Quantitative Values to Qualitative Factors in the Naval Electronics Program," *Naval Research Logistics Quarterly,* March 1959.

12. I. F. Stone, "McNamara and the Militarists," *New York Review of Books* 11(8): 7 November 1968, p. 5.

13. Ida R. Hoos, *Systems Analysis As a Technique for Solving Social Problems— A Realistic Overview* (Berkeley, Calif.: University of California, 1968), p. 56.

14. George Noel Kurilko, "Review of Charles E. Lindblom's The Intelligence of Democracy," *Journal of the American Institute of Planners* 34(3): May 1968, p. 198.

15. Braybrooke and Lindblom, *A Strategy of Decision,* p. 113.

16. Douglas Lee, "Requiem for Large-Scale Models," in *Journal of The American Institute of Planners,* May 1973, pp. 163–178.

## 11. Some Counter-Models and Their Utility

1. Vocabulary is often a problem here. Some writers have used as equivalent terms for "rationally": "comprehensively," "synoptically," "systematically," "analytically," . . .

2. David Braybrooke and Charles E. Lindblom, *A Strategy of Decision: Policy Evaluation as a Social Process* (New York: Free Press, 1963), p. 6.

3. Charles J. Hitch and Roland McKean, *The Economics of Defense in the Nuclear Age* (New York: Atheneum, 1967), p. 121.

4. The "naive priorities method," in Braybrooke and Lindblom, *A Strategy for Decision,* p. 7 and the "priorities approach," in Hitch and McKean, *The Economics of Defense,* p. 122; see also an example of the "requirements" approach in Robert J. Art, *The TFX Decision* (Boston: Little, Brown, 1968), pp. 32–33.

5. William Michelson, "Potential Candidates for the Designer's Paradise: A Social Analysis from a Nationwide Survey," *Social Forces,* 46(2), December 1967, pp. 190–197; and William Michelson, "Urban Sociology as an Aid to Urban Physical Development: Some Research Strategies," *American Institute of Planners* 34

(2), March 1968, pp. 105–108. See also note 2 chapter 9.

6. See the criticism of same in *Design Methods Group* Newsletter 2(11): November 1968, pp. 6–7.

7. Robert Venturi, Denise Scott Brown, Steven Izenour, *Learning from Las Vegas* (Cambridge: MIT Press, 1972), and John W. Cook and Heinrich Klotz, "Ugly is Beautiful," *Atlantic,* May 1973, pp. 33–43.

8. Braybrooke and Lindblom, *A Strategy of Decision,* p. vi.

9. Charles E. Lindblom, "The Science of Muddling Through," *Public Administration Review,* 1959.

10. Braybrooke and Lindblom, *A Strategy of Decision,* p. 84.

11. Ibid., pp. 88, 90–93.

12. Ibid., p. 86.

13. Ibid., p. 94.

14. Ibid., pp. 105–106.

15. Ibid., p. 106; Amitai Etzioni, *The Active Society* (New York: Free Press, 1968), pp. 268–273; Yehezkel Dror, "Muddling Through: 'Science' or Inertia," *Public Administration Review* 24, September 1964, pp. 297–298.

16. Herbert A. Simon, *Administrative Behavior* (New York: The Free Press, 1965, 2nd edition), pp. xxiii, 80–84, 241; and Herbert A. Simon, *Models of Man: Social and Rational* (New York: Wiley, 1957), pp. 204–205, 241. Herbert A. Simon, "Rational Choice and The Structure of The Environment," *Psychological Review* 63 (1956), pp. 129–138; reprinted in F. E. Emery, *Systems Thinking* (London: Penguin, 1970), pp. 214–229; see also "Cognitive Limits on Rationality," in James G. March and Herbert A. Simon, *Organizations* (New York: Wiley, 1958), pp. 136–171. For Etzioni's model, see: Amitai Etzioni, "Mixed Scanning: A 'Third' Approach to Decision Making," *Public Administration Review* 27(5), December 1967, pp. 385–392, and *The Active Society,* pp. 282–309.

17. Simon, *Models of Man,* p. 205.

18. Etzioni, *The Active Society,* p. 283.

## 12. Summary and Synthesis

1. George Alexander Steiner, *Managerial Long Range Planning* (New York: McGraw-Hill, 1963), p. 312.

2. H. A. Simon, *Administrative Behavior* (New York: The Free Press, 1965, 2nd edition), p. 5; and March and Simon, *Organizations,* pp. 190–194.

3. Ministry of Housing and Local Government, *The Future of Development Plans: A Report by the Planning Advisory Group* (London: HMSO, 1965). See also, J. D. Stewart, "The Administrative Structure of Planning," *Journal of the Town Planning Institute,* vol. 55, 1969, pp. 288–90. J. K. Friend and W. N. Jessop, *Local Government and Strategic Choice* (London: Tavistock, 1969), pp. 110–113. To the systems analyst, this approach is usually known as contingency planning: stressing "the need for sequential decision making, for improvisation, for hedging, and for adaptability." James R. Schlesinger, "The Changing Environment for Systems Analysis," in Stephen Enke, *Defense Management* (Englewood Cliffs, N.J.: Prentice-Hall, 1967), p. 109.

4. Peter L. Szanton, *Systems Problems in the City* (New York: The New York City-Rand Institute [P4821], April 1972), p. 9.

5. Donald N. Michael, "Ritualized Rationality and Arms Control," *Bulletin of*

*Atomic Scientists* 17(2), February 1961, p. 72. Robert Boguslow cites this in *The New Utopians: A Study of System Design and Social Change* (Englewood Cliffs, New Jersey: Prentice-Hall, 1965, p. 66) and adds an interesting comment:

> In many ways Michael is unfair in singling out logic and mathematics when discussing ritual modeling. Indeed, one might well use his phrase to describe psychological analyses that are ritualistically Freudian, Jungian, Adlerian, and so on, and are used to avoid the necessity for logical or empirical or mathematical analysis. The interesting point for our purposes is not the fact that logic or mathematics is used. Rather, it is that any method or technique—scientific, mathematical, psychological, or religious—may be invoked as a substitute for problem solving. It happens that in many contemporary decision-making circles, slide-rule bearing "whiz kids" have become fashionable. In other times and in other places it has been German epistemologists, Viennese psychiatrists, American combat pilots, or New England payroll-meeting businessmen who were the pets of fashion. . . .
>
> Ritual modeling is also useful, of course, as an alternative to action. A parliamentary device that has proved far more effective than the filibuster is the device of delaying action until a ritual modeling job has been completed. Frequently these efforts are called "studies," or "investigation," or even "research."

6. Clovis Heimsath, "The Systems Phenomenon," *Journal of the American Institute of Architects,* November 1973, p. 9.

7. Henri Lefebvre, Preface to Philippe Boudon, *Lived in Architecture: LeCorbusier's Pessac Revisited,* trans. Gerald Onn (Cambridge: MIT Press, 1972).

8. A certain utility in the use of analogy and metaphor in problem-solving has been assumed in this essay—to the obvious advantage of the organic theorists in architecture. The reader may question this premise; admittedly, it is a complex issue. Some insights that have informed and, I think, ultimately sustained my use of the premise can be found in: William K. Wimsatt, "Organic Form: Some Questions About a Metaphor," in G. S. Rousseau, *Organic Form,* pp. 61–81; Max Black, *Models and Metaphors* (Ithaca: Cornell University Press, 1954); Ernest Nagel, "The Role of Analogy," in *The Structure of Science* (New York: Harcourt, Brace & World, 1961), pp. 107–117, esp. p. 108; Michael Polanyi, *Personal Knowledge* (Chicago: University of Chicago Press, 1958), esp. pp. 57–58, 95–102; D. Berggram, "The Use and Abuse of Metaphor" (2 articles), in *The Review of Metaphysics,* 16 (1962/ 1963); Thomas Kuhn, "The Priority of Paradigms," in *The Structure of Scientific Revolutions* (Chicago: University of Chicago Press, 1962), pp. 43–51.

9. Regardless of specific values the analyst may or may not wish to accommodate, the suggestion is minimally for a larger perspective in perceiving the values which may influence the definition of a problem. On this point, Robert Boguslow is right in cautioning

> social scientists who insist upon a preoccupation with after-the-fact facts of life; intellectuals who concern themselves with rationalizations about society rather than with the realities of social issues; engineers and "hard" scientists who blithely deal with system data presented to them for analysis without wishing to ask questions about the broader context that lends more complete meaning to these data; and the beatniks who withdraw completely from all meaningful involvement in the affairs of their times from motives of indolence, ineptitude, or fear.
>
> The world, in a very real sense, belongs to those who know how to harness the dominant ethos of their times while escaping personal entrapment by that

ethos. In mid-twentieth-century Western civilization this ethos clearly includes a deep-seated reverence for organizational forms and perhaps an even deeper respect for the wonders that surround blinking lights and push-buttoned consoles. *The New Utopians,* p. 178.

10. Louis Kahn, quoted by Paul Goldberger in *New York Times,* 21 March 1974, p. 46.

11. Perhaps there is simply a diminution of the significance of the architects' "art" as perceived by modern society. A shift in values with postindustrialization leads to a trivializing of what once was art. See: Jacques Barzun, *The Use and Abuse of Art* (Princeton, Princeton University Press, 1974).

12. See: "PPB in a Political Context," Charles L. Schultze, *The Politics and Economics of Public Spending,* The H. Rowan Gaither Lectures in Systems Science (Washington: Brookings Institution, 1968), pp. 77–102.

# Bibliography

Listed below are suggestions for further reading. There is a limited selection from sources cited above in the notes as well as new material. The bibliography is grouped according to the subject matter of the three parts of the book—with some inevitable overlap of content.

## 1 The Systems Idea

Abrams, Meyer H. *The Mirror and The Lamp: Romantic Theory and The Critical Tradition.* New York: Norton, 1958.

ABT Associates. *Applications of Systems Analysis Models.* Washington: National Aeronautics & Space Administration, 1968.

Ackoff, Russell L. and Emery, Fred E. *On Purposeful Systems.* New York: Aldine-Atherton, 1972.

Adams, Richard P. "Architecture and the Romantic Tradition: Coleridge to Wright." *American Quarterly* 9(1): Spring 1957, pp. 46–62.

Adelman, Marvin. "The Systems Approach." *Ekistics,* vol. 23, June 1967, pp. 311–315.

Alexander, Christopher. *Notes on the Synthesis of Form.* Cambridge: Harvard, 1964.

Behrendt, Walter Curt. *Modern Building: Its Nature, Problems, and Forms.* New York: Harcourt Brace, 1937.

Building Performance Research Unit, *Building Performance.* New York: Wiley, 1972.

Churchman, C. West. *The Systems Approach.* New York: Dell, 1968.

Cleland, David I. and King, William R. *Systems Analysis and Project Management.* New York: McGraw-Hill, 1968.

de Zurko, Edward R. *Origins of Functionalist Theory.* New York: Columbia, 1957.

Egbert, Donald Drew. "The Idea of Organic Expression and American Architecture." *Evolutionary Thought in America.* Persons, Stow, ed. New Haven: Yale, 1950, pp. 336–396.

Emery, F. E., ed. *Systems Thinking.* Baltimore: Penguin Books, 1969.

Fisk, George, ed. *The Analysis of Business Systems.* Lund, Sweden: CWK Gleerup, 1967.

Gilbert, Katherine. "Clean and Organic: A Study in Architectural Semantics." *Journal of the Society of Architectural Historians* 10(3): October 1951, pp. 3–7.

Herbert, Gilbert. "Form and Function: A Study of Frank Lloyd Wright's Theory of Organic Architecture." *South African Architectural Record* 44(9): September 1959, pp. 23–29.

————. *The Synthetic Vision of Walter Gropius.* Johannesburg: Witwatersrand University Press, 1959.

————. "The Organic Analogy in Town Planning." *Journal of the American Institute of Planners* 29(3): August 1963, pp. 198–209.

Herrey, Hermann et al. "An Organic Theory of City Planning." *Architectural Forum* 80(4): April 1944, pp. 133–140.

Hitch, Charles J. and McKean, Roland. *The Economics of Defense in The Nuclear Age.* New York: Atheneum, 1967.

Jones, J. Christopher. *Design Methods: Seeds of Human Futures.* New York: Wiley-Interscience, 1970.

Laszlo, Ervin. *The Systems View of the World.* New York: Braziller, 1972.

Lee, Alec M. *Systems Analysis Frameworks.* New York: Wiley, 1970.

Lord, Catherine. "Organic Unity Reconsidered." *Journal of Aesthetics and Art Criticism,* vol. 52, 1964, pp. 263–268.

Martindale, Don. *The Nature and Types of Sociological Theory.* Boston: Houghton Mifflin, 1960.

Modisette, Eldon L. "The Legitimization of Modern American Architecture." *Journal of Aesthetics and Art Criticism* 20(3): Spring 1962, pp. 251–261.

Nitschke, Gunther. "The Metabolists of Japan." *Architectural Design* 34(10): Oct. 1964, pp. 509–524.

Pepper, Stephen C. *The Basis of Criticism in The Arts.* Cambridge: Harvard, 1963.

Quade, E. S. and W. I. Boucher, eds. *Systems Analysis and Policy Planning.* New York: American Elsevier, 1969, pp. 345–363.

Rotival, Maurice E. H. "Une Nouvelle Methode De Planification." *Urbanisme,* no. 76, 1963, pp. 8–21.

Rousseau, G. S., ed. *Organic Form: The Life of an Idea.* London: Routledge and Kegan Paul, 1972.

Sayles, Leonard R. and Margaret K. Chandler. *Managing Large Systems: Organizations for the Future.* New York: Harper and Row, 1971.

Shuchman, Abe, ed. *Scientific Decision Making in Business.* New York: Holt, Rinehart and Winston, 1963.

Sullivan, Louis H. *Kindergarten Chats and Other Writings.* New York: George Wittenborn, 1947.

"Systems Approach to Problem-Solving." *Public Management,* Feb. 1969, entire issue.

von Bertalanffy, Ludwig. *Problems of Life.* New York: Wiley, 1952.

————. "General Systems Theory: A Critical Review." *General Systems Yearbook,* vol. 7, 1962.

————. *General System Theory.* New York: Braziller, 1968.

Young, O. R. "A Survey of General Systems Theory." *General Systems Yearbook,* vol. 9, 1964, pp. 61–80.

Zucker, Paul. "The Paradox of Architectural Theories at the Beginning of the Modern Movement." *Journal of the Society of Architectural Historians* 10(3), Oct. 1951, pp. 8–14.

Zevi, Bruno. *Towards an Organic Architecture.* London: Faber and Faber, 1950.


## 2   Use of the Systems Idea

Ackoff, Russell L. *Scientific Method.* New York: Wiley, 1968.

Aguilar, Rodolfo J. *Systems Analysis and Design in Engineering, Architecture, Construction and Planning.* Englewood Cliffs, N.J.: Prentice-Hall, 1973.

Alfandary-Alexander, Mark, ed. *Analysis for Planning-Programming-Budgeting.* Washington: Washington Operations Research Council, 1968.

Blumstein, Alfred, Kamrass, Murray, and Weiss, Armand B., eds. *Systems Analysis for Social Problems.* Washington: Washington Operations Research Council, 1970.

Chadwick, George. *A System View of Planning.* New York: Pergamon Press, 1971.

Churchman, C. West, Ackoff, R., and Arnoff, E. L. *Introduction to Operations Research.* New York: Wiley, 1968.

Cohn, Elchanan. *Public Expenditure Analysis.* Lexington, Mass.: Lexington Books, 1972.

Dietz, Albert G. H. and Cutler, Laurence S., eds. *Industrialized Building Systems for Housing.* Cambridge: MIT, 1971.

Dorfman, Robert, ed. *Measuring Benefits of Government Investments.* Washington: The Brookings Institution, 1965.

Fisher, Gene H. *Cost Considerations in Systems Analysis.* New York: American Elsevier, 1971.

Flagle, Charles D. et al. *Operations Research and Systems Engineering.* Baltimore: Johns Hopkins, 1960.

Friend, J. K. and Jessop, W. N. *Local Government and Strategic Choice.* Beverly Hills: Sage, 1969.

Hassid, Sami. "Systems of Judgment of Architectural Design." *New Building Research.* Washington: National Academy of Sciences and the National Research Council, 1961.

Haveman, Robert H. and Margolis, Julius, eds. *Public Expenditures and Policy Analysis.* Chicago: Markham, 1970.

Hemmens, George C., ed. *Urban Development Models.* Washington: Highway Research Board, Special Report no. 97, 1968.

Hitch, Charles J. and McKean, Roland. *The Economics of Defense in the Nuclear Age.* New York: Atheneum, 1967.

Jones, Barclay G. "Design Process and Decision Theory." *Teaching of Architecture.* Wiffen, Marcus, ed. Washington: The American Institute of Architects, 1964.

Kendall, M. G., ed. *Cost-Benefit Analysis.* New York: American Elsevier, 1971.

Kilbridge, Maurice D. et al. "A Conceptual Framework for Urban Planning Models." *Management Science* 15(6): Feb. 1969, pp. 246–266.

Kraemer, Kenneth L., Mitchel, William H., Weiner, Myron E., and Dial, O. E. *Integrated Municipal Information Systems.* New York: Praeger, 1973.

Kraemer, Kenneth L. *Policy Analysis in Local Government: A Systems Approach to Decision-Making.* Washington: International City Management Association, 1973, p. 31.

**159**

Layard, Richard, ed. *Cost-Benefit Analysis.* Baltimore: Penguin Books, 1972.

Levin, Richard I. and Lamone, Rudolph P., eds. *Quantitative Disciplines in Management Decisions.* Belmont, Calif.: Dickenson, 1969.

Lichfield, Nathaniel. "Evaluation Methodology of Urban and Regional Plans: A Review." *Regional Studies,* vol. 4, 1970, pp. 151–165.

———. "Cost-Benefit Analysis in Planning: A Critique of The Roskill Commission." *Regional Studies,* vol. 5, 1971, pp. 157–183.

Lyden, Fremont J. and Miller, Ernest G. *Planning Programming Budgeting—A Systems Approach to Management.* 2nd edition. Chicago: Markham, 1972.

Margolis, Julius, ed. *The Public Economy of Urban Communities.* Washington: Resources for the Future, 1965.

———., ed. *The Analysis of Public Output.* New York: National Bureau of Economic Research, 1970.

McLoughlin, B. J. *Regional and Urban Planning: A Systems Approach.* London: Faber and Faber, 1969.

McKean, Roland N. *Public Spending.* New York: McGraw-Hill, 1968.

Mishan, E. J. *Elements of Cost-Benefit Analysis.* London: George Allen and Unwin, 1972.

Moore, Gary T., ed. *Emerging Methods in Environmental Design and Planning.* Cambridge: MIT, 1970.

Morse, Philip M. and Bacon, L. W., eds. *Operations Research for Public Systems.* Cambridge: MIT, 1967.

Pack, Janet Rothenberg et al. *The Use of Urban Models in Urban Policy Making.* Philadelphia: Univ. of Pennsylvania/Fels Center of Government, 1974.

Porter, William A. *Modern Foundations of Systems Engineering.* New York: Macmillan, 1966.

Proshansky, Harold M., Ittelson, William H., and Rivlin, Leanne G., eds. *Environmental Psychology: Man and His Physical Setting.* New York: Holt, 1970.

Reif, Benjamin. *Models in Urban and Regional Planning.* New York: Intertext, 1973.

Rivlin, Alice M. *Systematic Thinking for Social Action.* Washington: Brookings Institution, 1971.

Robinson, Ira M., ed. *Decision-Making in Urban Planning: An Introduction to New Methodologies.* Beverly Hills: Sage, 1972.

Rothenberg, Jerome. *Economic Evaluation of Urban Renewal: Conceptual Foundation of Benefit-Cost Analysis.* Washington: Brookings Institution, 1967.

Schaller, Howard G. *Public Expenditure Decisions in the Urban Community.* Washington: Resources for the Future, 1963.

Schmid, Thomas and Testa, Carlo. *Systems Building: An International Survey of Methods.* New York: Praeger, 1969.

Sweet, David C. *Models of Urban Structure.* Lexington, Mass.: D. C. Heath, 1972.

Terner, Ian Donald and Turner, John F. C. *Industrialized Housing.* Washington: U.S. Agency for International Development and Department of Housing and Urban Development, 1972.

Ward, R. A. *Operational Research in Local Government.* London: George Allen and Unwin, 1964.

Williams, Walter. *Social Policy Research and Analysis.* New York: American Elsevier, 1971.

Wilson, A. G. *Urban and Regional Models in Geography and Planning.* New York: Wiley, 1974.

Andreski, Stanislav. *Social Sciences as Sorcery.* London: Andre Deutsch, 1973.

Berkman, Herman G. "The Scope of Scientific Technique and Information Technology in Metropolitan Area Analysis," *Annals,* Monograph VII. Philadelphia: American Academy of Political and Social Science, May 1967.

Boguslaw, Robert. *The New Utopians: A Study of System Design and Social Change.* Englewood Cliffs, New Jersey: Prentice-Hall, 1965.

Braybrooke, David and Lindblom, Charles E. *A Strategy of Decision: Policy Evaluation as a Social Process.* New York: Free Press, 1963.

Brewer, Garry D. *Politicians, Bureaucrats, and the Consultant: A Critique of Urban Problem Solving.* New York: Basic Books, 1973.

Chase, Jr., Samuel B., ed. *Problems in Public Expenditure Analysis.* Washington: The Brookings Institution, 1968.

Diesing, Paul. *Reason in Society.* Urbana: Univ. of Illinois, 1962.

Dror, Yehezkel. *Public Policy-Making Reexamined.* San Francisco: Chandler, 1968.

Dyckman, J. W. "The Practical Uses of Planning Theory." *Journal of the American Institute of Planners 35.* September 1969.

Etzioni, Amitai. *The Active Society.* New York: Free Press, 1968.

Friedrich, Carl J. *The Public Interest.* New York: Atherton, 1966.

Glazer, Nathan. "The Limits of Social Policy." *Commentary* 52(3): September 1971, pp. 51–58.

Haveman, Robert H. and Margolis, Julius. *Public Expenditures and Policy Analysis.* Chicago: Markham, 1970.

Heilbroner, Robert L. *An Inquiry into the Human Prospect.* New York: Norton, 1974.

Heimer, Olaf. *Social Technology.* New York: Basic Books, 1966.

Hoos, Ida R. *Systems Analysis in Public Policy: A Critique.* Berkeley: Univ. of California, 1972.

Jackson, Anthony. *The Politics of Architecture.* Toronto: Univ. of Toronto, 1970.

Kuenzlen, Martin. *Playing Urban Games: The Systems Approach to Planning.* Boston: The i Press, 1972.

Lindblom, Charles E. *The Policy-Making Process.* Englewood Cliffs, N.J.: Prentice-Hall, 1968.

Mailick, Sidney and Van Ness, Edward H., eds. *Concepts and Issues in Administrative Behavior.* Englewood Cliffs, N.J.: Prentice-Hall, 1962.

Meyerson, Martin and Banfield, Edward C. *Politics, Planning and the Public Interest.* Glencoe: The Free Press, 1955.

Miller, James R. *Professional Decision-Making: A Procedure for Evaluating Complex Alternatives.* New York: Praeger, 1970.

Mitchell, Joyce M. and Mitchell, William C. *Policy-Making and Human Welfare.* Chicago: Rand McNally, 1969.

Moynihan, Daniel P. *Coping.* New York: Random House, 1974.

Nelkin, Dorothy. *The Politics of Housing Innovation.* Ithaca: Cornell, 1971.

Niskanen, William A., Jr. *Bureaucracy and Representative Government.* Chicago: Aldine-Atherton, 1971.

Pressman, Jeffrey L. and Wildavsky, Aaron B. *Implementation.* Berkeley: Univ. of California, 1973.

Raymond, George M. "Simulation vs. Reality." *Urban Planning in Transition.* Ernest Erber, ed. New York: Grossman, 1970.

Roberts, Walter Orr. "Science, a Wellspring of Our Discontent." *American Scholar*, vol. 36, Spring 1967.

Schoeck, Helmut and Wiggens, James W. eds. *Scientism and Values.* Princeton: Van Nostrand, 1960.

Schultze, Charles L. *The Politics and Economics of Public Spending.* Washington: Brookings Institution, 1968.

Stein, Bruno and Miller, S. M., eds. *Incentives and Planning in Social Policy.* Chicago: Aldine, 1973.

Suchman, Edward A. *Evaluative Research.* New York: Russell Sage Foundation, 1967.

Vickers, Geoffrey. *The Art of Judgment.* London: Chapman and Hall, 1965.

Weisskopf, Walter A. *Alienation and Economics.* New York: Dutton, 1971.

Whitehead, Clay Thomas. *Uses and Limitations of Systems Analysis.* Santa Monica: Rand Corporation, 1967.

# Index

Jonas, Walter, 56

Kahn, Louis, 54, 95
*Kindergarten Chats* (Sullivan), 34–35
kitchens as subsystems, 24
Koontz, Harold, 20
Kraemer, Kenneth, 14
Krick, Edward, 17, 24

labor, skilled, 59, 65, 68
labor, unskilled, 65
labor unions, 66–67, 69
Landrum-Griffin Act, 66
land use, 86, **24**
land values, 86
Laquer, Walter, 30
Las Vegas technique, 83, 93
Le Corbusier (Charles Jeanneret), 2, 31,
    42, 53, 54, 56, 60, 66, 69, 71, 93, 94
Lee, Douglas, 82
Lefebvre, Henri, 93, 94
legal costs, 47
legal rationality, 15
legislation, housing, 68
Leighton, Alexander H., 13
Le Nôtre, André, 2
LePlay, Frédéric, 43
Levin, Earl, 7
Lichfield, Nathaniel, 51–52, 54
Lindblom, Charles E., 81–87
loans, 46
Long Island, New York, 78
Louis XIV of France, 2
Lower East Side, New York, 6
Lower Manhattan Expressway, 6, 78

machines vs. brains, 39
MacNamara, Robert, 25, 27
Maki, Fumihiko, 56
Mannheim, Karl, 5
masonry, 61
mass production, 59, 62, 64
materials, construction, 1, 60, 61, 63, 68
McKean, Roland, 8, 9, 82
mechanism, 33
medieval architecture, 1, 2, 29, 33, 35,
    53
Metabolist Group, 42, 95
metaphysics, 28, 37, 40
Michael, D. N., 91
Michelangelo, 38
military systems, 3, 21–24, 27, 70–72, 74,
    79
Miller, James, 12
mobile homes, 61, 65, **46**
mobility, social, cultural, and geographi-
    cal, 71
models, planning, 7, 18, 39, 41, 42, 54,
    55–58, 72–73, 79, 82, 85, **5, 21–25,
    54**
    definition of, 18

predictive, 18
procedural, 19–20
process, 20, 24
Model T Fords, 64
modern architecture, 37
Modissette, Eldon L., 40
modules, housing, 60–61, 64, 66–67, **41,
    42**
money flow, *see* cash flow
moral values, 80
Morris, William, 29, 31
Mumford, Lewis, 29, 31, 35, 36, 41, 42, 43
Myrdal, Gunnar, 68

nature:
    and architecture, 37, 41
    and planning, 31, 34–35, 42–43
naturalism, 41
neighborhoods, 72
    as subsystems, 11, 28, 31
    as systems, 11
    *see also* communities
nervous systems, human, 39
Neutra, Richard, 38, 42
New York City, 6–7, 78
Nietzsche, Friedrich, 32
Nisbet, Robert, 29, 30
Novalis (Friedrich von Hardenberg), 30
Nuremburg, Castle of, 33

O'Donnell, C. J., 20
*On Growth and Form* (Thompson), 41
Operation Breakthrough, 63
optimal performances, 87
    *see also* suboptimization
organic architecture, 28–38, 91–92, 94–
    95
    definition of, 35
organicism, 33, 35–36, 38, 42, 43
organisms, living, as systems, 27–28, 31,
    38–42
organisms, social, 40
organizations:
    definition of, 6
    as systems, 5–6
outputs, from systems, 10–11, 14, 16, 17

panel systems, prebuilt, 60, 61, **33, 34**
Park, Robert E., 29
Parsons, Talcott, 12
Paxton, Joseph, 60
Pentagon, *see* Defense Department, U.S.
"Philadelphia Door" case, 66
philosophy, 8, 37
physical forms:
    in architecture, 28, 38, 41, 44
    of cities, 4, 7
    of systems, 10, 21–23
    *see also* "form follows function"
planning balance sheets, 51–52, 54, **18**